Analysis of Reaction and Transport Processes in Zinc Air Batteries

Daniel Schröder

Analysis of Reaction and Transport Processes in Zinc Air Batteries

Daniel Schröder
Gießen, Deutschland

Genehmigte Dissertation an der TU Braunschweig, Fakultät für Maschinenbau, 2015

ISBN 978-3-658-12290-4   ISBN 978-3-658-12291-1 (eBook)
DOI 10.1007/978-3-658-12291-1

Library of Congress Control Number: 2015959488

Springer Vieweg
© Springer Fachmedien Wiesbaden 2016

Printed on acid-free paper

Springer Vieweg is a brand of Springer Fachmedien Wiesbaden
Springer Fachmedien Wiesbaden is part of Springer Science+Business Media
(www.springer.com)

Analysis of Reaction and Transport Processes in Zinc Air Batteries

Von der Fakultät für Maschinenbau
der Technischen Universität Carolo-Wilhelmina zu
Braunschweig

zur Erlangung der Würde

eines Doktor-Ingenieurs (Dr.-Ing.)

genehmigte Dissertation

von: Dipl.-Ing. Daniel Schröder
aus (Geburtsort): Waren (Müritz)

eingereicht am: 19.02.2015
mündliche Prüfung am: 17.07.2015

Gutachterinnen / Gutachter:
Prof. Dr.-Ing. Ulrike Krewer
Prof. Dr.-Ing. Thomas Turek
Prof. Dr.-Ing. Arno Kwade

2015

Acknowledgments

The research for this thesis was conducted during the last four years partly at the Otto-Hahn Research group for Portable Energy Systems at the Otto-von-Guericke University Magdeburg and at the Max-Planck-Institute for Dynamics of Complex Technical Systems in Magdeburg, and at the Institute of Energy and Process Systems Engineering at the TU Braunschweig. During this period, I had the pleasure to work alongside many people and I want to express my gratitude to them at this point.

First of all, I thank Prof. Dr.-Ing. Ulrike Krewer who gave me the opportunity to work on the topic of zinc air batteries, and who guided and encouraged me throughout the entire time. Her enormous attention to detail and critical thinking helped me to shape the thesis, with all the minor steps within it, to the current state. Furthermore, I thank Prof. Dr.-Ing. Thomas Turek from the TU Clausthal for the interest in my thesis, the helpful scientific discussions at various conferences, and for assessing the thesis. I also thank Prof. Dr.-Ing. Arno Kwade for helpful discussions at various conferences and project meetings, and for chairing the doctoral committee.

A great amount of gratitude belongs to my closest colleagues in Magdeburg (and later Braunschweig), namely Qing Mao, Prashant Khadke, Maik Kraus, Nebojsa Korica, Youngseung Na and Christine Weinzierl. It was always a pleasure to discuss with you about the fundamentals of research and life. I am strongly indebted to you for several discussions that helped me to shape the experimental and model-based parts of this thesis. Furthermore, I want to thank Thomas Khadyk, Christian Oettel, Christoph Hertel, Peter Heidebrecht, Mathias Pfafferodt and Federico Zenith of the Max-Planck-

Institute for sharing their thoughts on research and life in general at the daily lunch breaks.

I appreciate the help of my colleagues Niels Brinkmeier, Paul Alps, Andreas Hauschke and Sebastian Stenger at the institute in Braunschweig; each of you was essential to help me to settle and feel welcome in Braunschweig. I also thank my colleagues in the battery research group in Braunschweig, Georg Lenze, Fridolin Röder, Angelica Staeck and Nan Lin. I am indebted to Horst Müller for the numerous discussions/lessons about numerical methods, teaching, and accounting. I also thank Victor Emenike and Georg Lenze for the fruitful discussions (topical and non-topical) in our office.

In addition, at the Max-Planck-Institute Magdeburg I would like to thank the library team, the members of the mechanical and the electrical workshop, and the lab assistants Bianka Stein, Markus Ikert and Jessica Bunge. Thanks also belong to Ina Schunke, Nina Böge, Wilfried Janßen, Sergej Maserow and Uwe Herrmann, since they provided me with technical and non-technical advice at all time at the TU Braunschweig. Their contribution is often under-appreciated but crucial for good research.

I also thank Ingo Manke and especially Tobias Arlt of the Helmholtz-Zentrum Berlin, who contacted me for the joint research on X-ray tomography of zinc air batteries. I appreciate the numerous hours to plan, conduct and analyze the X-ray measurements together. I thank them for the various discussions that led to our joint publications, and the X-ray images provided for parts of this thesis.

Moreover, I thank the students/coworkers Ilona Heidenreich, Thomas Gebken, Markus Pollack, Jan-Robert Schwarz, thereof especially Becca McClain, Vincent Laue, Michael König, and Neeraj Nitin Sinai Borker for the joyful and fruitful research together. I also acknowledge Becca McClain, Prashant Khadke, Christine Weinzierl, Youngseung Na, Tobias Arlt, and Victor Emenike for their critical review and proofreading of the manuscript of this thesis.

Finally, I want thank my family and Carla, who always supported and encouraged me the most.

Contents

List of Figures

List of Tables

List of Abbreviation

Notation	Description
CL	catalyst layer
DL	double layer
EOD	electro-osmotic (water)drag
GDE	gas diffusion electrode
GDL	gas diffusion layer
HER	hydrogen evolution reaction
OCP	open circuit potential
ODE	ordinary differential equation
OER	oxygen evolution reaction
ORR	oxygen reduction reaction
PPS	Polyphenylene sulfide
PTFE	Polytetrafluoroethylenc or Teflon
PVC	Polyvinyl chloride
SHE	standard hydrogen potential
SOD	state-of-discharge
ZAB	zinc air battery

List of Symbol

Term Symbol	Description	Unit
A	cross-sectional area	dm^2
a	species activity on molar basis	–
B	term to summarize other terms	variable unit
b	empiric parameter	variable unit
c	species molar concentration	$mol \cdot dm^{-3}$
C_{Ah}	actual battery capacity	$mA \cdot h$
$C_{Ah,initial}$	initial battery capacity according to grams of active material Zn; for example 819 mA·h for one gram of Zn	$mA \cdot h$
$\bar{c}$	species molar concentration directly at the interface of gas and liquid	$mol \cdot dm^{-3}$

– continued –

$\check{c}$	average species molar concentration, approximated as mean value of the indicated species concentrations	$\mathrm{mol \cdot dm^{-3}}$
C_{DL}	double layer capacitance	$\mathrm{F \cdot dm^{-2}}$
$\tilde{c}$	species molal concentration	$\mathrm{mol \cdot kg^{-1}}$
D	diffusion coefficient	$\mathrm{dm^2 \cdot s^{-1}}$
d	diameter	dm
E	cell potential	V
En	enhancement factor for the absorption process	–
F	volumetric flow rate	$\mathrm{dm^3 \cdot s^{-1}}$
$\mathbf{F}$	Faraday's constant, 96487	$\mathrm{A \cdot s \cdot mol^{-1}}$
f	frequency	Hz
H	Henry's law constant	$\mathrm{mol \cdot (dm^3 \cdot Pa)^{-1}}$
Ha	Hatta modulus	–
I	current	A

– continued –

i	current density	$A \cdot dm^{-2}$
i_{limit}	limiting current density	$A \cdot dm^{-2}$
$\bar{I}$	intensity of radiation exiting a sample, i.e. transmitted	$J \cdot dm^{-2} \cdot s^{-1}$
$\bar{I}_0$	intensity of radiation entering a sample	$J \cdot dm^{-2} \cdot s^{-1}$
J	molar flow rate	$mol \cdot s^{-1}$
j	imaginary unit for the use in a complex number	–
k	reaction rate constant	variable unit
k_B	Boltzmann constant	$m^2 \cdot kg^{-2} \cdot s^{-2} \cdot K^{-1}$
k_d	dimensionless water drag coefficient	–
$k_{CO_2}^{electrolyte}$	mass transfer coefficient for the absorption of CO_2 into the electrolyte	$dm \cdot s^{-1}$
$g\text{-}k,\ o,\ m,\ n$	counter/control variables indicating respective species	–
l	length	dm

– continued –

m^*	exponent in the Bruggeman correlation	–
m	mass	g
N	molar flux	$mol \cdot s^{-1} \cdot dm^{-2}$
n	species molar amount	mol
n	number of a volume element	–
p	pressure or partial pressure	Pa
Q	source or sink term	$mol \cdot s^{-1} \cdot dm^{-3}$
$\mathbf{R}$	universal gas constant, 8.3143	$J \cdot (mol \cdot K)^{-1}$
r	reaction rate	$mol \cdot s^{-1}$
R	ohmic resistance	Ω
RH	relative humidity	–
R_{ion}	effective hydrodynamic radius of an ion; Stokes radius	m
R^*	area specific impedance, respectively internal battery resistance	$\Omega \cdot dm^2$

– continued –

S_{ionic}	ionic strength of an electrolyte solution	$\mathrm{mol \cdot dm^{-3}}$
T	temperature	K
t	time	s
t_k	transference number of ion k	$-$
V	volume	$\mathrm{dm^3}$
V_{l}	volume in the liquid filled part of the gas diffusion layer in the air electrode model	$\mathrm{dm^{-3}}$
$\tilde{V}$	partial molar volume	$\mathrm{dm^3 \cdot mol^{-1}}$
w	mass fraction	$-$
x	molar fraction	$-$
x	space coordinate	dm
y	molar fraction in the gas phase	$-$
Z	area specific impedance	$\Omega \cdot \mathrm{dm^2}$
z	charge number of ion	$-$

– continued –

z_e	number of electrons participating in the electrochemical reaction indicated	–
$\hat{Z}$	impedance	Ω

Greek symbol

α	transfer/symmetry coefficient	–
χ	relative water content, ratio between actual molar amount of water and initial molar amount of water	–
δ	thickness, e.g. of electrode or separator	dm
$\Delta\Phi$	potential gradient between electrolyte phases in air and zinc electrode	$V \cdot dm^{-1}$
ε	porosity	–
η	electrode overpotential	V
η^{vis}	dynamic viscosity of the electrolyte	$Pa \cdot s^{-1}$
Γ	indicating an interface	–

– continued –

γ	species activity coefficient on molar basis	–
ι	indicating a position in the spatial domain after discretization	dm
κ	ionic conductivity	$S \cdot dm^{-1}$
λ_{O_2}	oxygen excess factor	–
μ	linear attenuation coefficient	dm^{-1}
μ^m	mass attenuation coefficient	$dm^2 \cdot g^{-1}$
Ω	indicating a domain filled with gas or liquid phase on the space coordinate	–
ϕ	phase angle	rad or °
ρ	volumetric mass density of a material	$g \cdot dm^{-3}$
σ	electric conductivity	$S \cdot dm^{-1}$
τ	tortuosity	–

Subscript

atm	indicating atmospheric pressure

– continued –

backw	indicating a backward reaction
electrode	indicating the entire electrode
electrolyte	indicating the electrolyte, i.e. liquid phase
film	indicating the liquid film thickness
forw	indicating a forward reaction
free	indicating the accessible area for species transport
g	indicating the gas phase in the air electrode model
gas	indicating supplied air or pure oxygen for the active operation
half	indicating the needed sample thickness to reduce the X-ray intensity of a beam by half
imag	indicating the imaginary part of a complex number
ionic	indicating solely the contribution of ionic resistance

– continued –

l	indicating the liquid phase in the air electrode model
lg	indicating the interface between gas and liquid in the air electrode model
measured	indicating an experimentally determined value
membrane	indicating the polymer-backbone of a separator
ohmic	indicating solely the contribution of ohmic resistance
permeation	indicating a permeating species, for instance in a separator
pulse	indicating the time for an applied pulse-current
real	indicating the real part of a complex number
solid	indicating zinc and zinc oxide particles, i.e. solid phase

– continued –

void	free space, not filled with solid material

Superscript

air	indicating the air electrode, which is the cathode of the zinc air battery
Amp	amplitude
a	anodic
*	indicating a dissolved species in a liquid
bulk	referring to the bulk of a solution or gas phase
c	cathodic
cell	indicating the entire zinc air battery
conv	convective contribution to a flux
diff	related to diffusion due to concentration gradient
drag	indicating electro-osmotic drag along with water

– **continued** –

eff	indicating an effective value of a standard, which is corrected for a special case
env	referring to a location just outside the battery; indicating the environment or surrounding of the battery
GDL	referring to the gas diffusion layer of the air electrode
interface	indicating the interface between gas phase and liquid phase at the air electrode
max	referring to the maximal value
mig	migration of a charged species due to potential gradient
ref	referring to a concentration of one mole per liter
sat	indicating the saturation of a species in a solvent
sep	indicating the battery separator

– continued –

sol   referring to the solubility con-
                    centration of a salt in water

stoic   indicating the stoichiometric
                    amount needed for a reaction

total   indicating the sum of moles of
                    a species at the beginning of a
                    simulation or experiment

0   open circuit; zero current condi-
                    tion

zinc   indicating the zinc electrode,
                    which is the anode of the zinc
                    air battery

Abstract

With renewable energies in mind, batteries might be applied in the upcoming years to store energy from intermittent energy sources, such as wind power or solar power. Thereof, the zinc air battery is one promising next-generation battery, for which only limited charge/discharge cycle numbers are reported. Hence, the main research focus for zinc air batteries is on improving the materials applied to overcome this drawback. In this thesis, another approach is used: A detailed analysis of the reaction and transport processes that aims to gain a better understanding of the operation of zinc air batteries and to understand how the cycle numbers can be improved. This is realized by means of a combined experimental and model-based analysis. The analysis helps to answer which battery composition and which air-composition should be adjusted to maintain stable and efficient charge/discharge cycling for zinc air batteries.

In the first part of this thesis, electrochemical investigations and X-ray transmission tomography are applied on button cell zinc air batteries and in-house set-ups. Thereby, various battery compositions (electrolyte, catalyst) and operation parameters (state-of-discharge, discharge current density) are investigated to qualitatively analyze their impact on the processes in the battery. Furthermore, a model-based analysis of the air electrode is presented to extend the experimental analysis by simulations of the overpotential and the oxygen distribution. The experimental results reveal that the volume of the zinc electrode expands during discharge, which in turn can cause flooding of the air electrode with liquid electrolyte. The model-based analysis clarifies that flooding can lead to a shortage of oxygen in the air electrode, and thus to battery failure.

In the second part of this thesis, the challenges that arise from the fact that zinc air batteries are half-open to the surrounding are addressed. A mathematical model of an electrically rechargeable zinc air battery is introduced, and subsequently modified to account for the impact of relative humidity, active operation, carbon dioxide and oxygen. The simulation results imply that performance and cycle numbers for zinc air batteries can be improved, if they are operated with pure oxygen and low amounts of carbon dioxide impurities. Depending on the application and the relative humidity in the surrounding, either the relative humidity or the electrolyte composition should be controlled or adjusted to avoid water loss or gain for the battery, and thus to prevent battery failure. However, additional equipment, such as blowers, controllers or filters, might be needed, which would diminish the practical energy density.

The experimental and model-based techniques used in this thesis complement each other well and yield a comprehensive understanding of zinc air batteries. The methods applied are adaptable and can potentially be applied to improve other metal-air batteries.

Kurzfassung

Vor dem Hintergrund der Energiewende bieten sich Batterien als möglicher Energiespeicher für die nicht immer gleichmäßig bereitgestellte Energie von Solar- und Windkraft an. Ein besonders vielversprechender Batterietyp ist die elektrisch wiederaufladbare Zink-Luftsauerstoff-Batterie. Für diesen Batterietyp sind jedoch gegenwärtig nur geringe Lade- und Entladezyklen erreichbar. Aus diesem Grund werden die eingesetzten Materialien zurzeit intensiv erforscht und gezielt variiert. In der vorliegenden Arbeit wird ein anderer Ansatz verfolgt: die Arbeit zielt darauf ab, die Reaktions- und Transportvorgänge in Zink-Luftsauerstoff-Batterien und deren Betriebsverhalten genauer zu analysieren, um die vorhandenen Limitierungen zu verstehen und ihnen entgegenwirken zu können. Hierzu wird eine Kombination aus experimentellen und modellbasierten Methoden eingesetzt. Mit der Arbeit kann beantwortet werden, aus welchen Komponenten Zink-Luftsauerstoff-Batterien zusammengesetzt sein sollten und welche Mengen an Nebenbestandteilen außer Sauerstoff in der Umgebungsluft zulässig sind, um eine größere Anzahl an Lade- und Entladezyklen als bisher möglich zu erreichen.

Im ersten Teil der Arbeit werden zunächst kommerziell erhältliche Knopfzellenbatterien und eigens konzipierte Batterien mit Hilfe von elektrochemischen Analysen und mittels Transmission-Röntgen-Computertomographie untersucht, um den Einfluss verschiedener Batterie-Komponenten (Elektrolyt, Kathoden-Katalysator) und verschiedener Betriebsbedingungen (Ladezustand, Entladestromdichte) auf die Batterie qualitativ zu beschreiben. Diese experimentellen Untersuchungen zeigen, dass das Volumen des Feststoffes in der Anode während des Entladevorgangs signifikant vergrößert und der

Flüssig-Elektrolyt aus diesem Grund zur Kathode transportiert wird, deren Poren dadurch geflutet werden. Das Auftreten dieses Effektes kann zu einem verfrühten Versagen der Zink-Luftsauerstoff-Batterie führen. Anschließend wird eine modellbasierte Analyse der Kathode der Batterie vorgestellt. Mit einem eindimensionalem Modell werden die Sauerstoffverteilung in der Kathode und die daraus resultierende Überspannung berechnet. Mit dieser Analyse werden die Folgen des zuvor festgestellten Effektes der Kathoden-Flutung für die Leistungsfähigkeit der Batterie näher untersucht.

Der zweite Teil der Arbeit beschäftigt sich mit den Herausforderungen, die sich aus der halb-offenen Bauweise der Zink-Luftsauerstoff-Batterien zur Umgebungsluft ergeben. Hierzu wird ein flexibel einsetzbares und erweiterbares Basismodell einer elektrisch wiederaufladbaren Zink-Luft-sauerstoff-Batterie aufgestellt und anschließend erweitert, um mittels Simulationen dezidiert die Einflüsse folgender Parameter zu analysieren: relative Luftfeuchte der Umgebung, Überschuss an aktiv zugeführtem Sauerstoff an der Kathode, Kohlendioxid-Gehalt und Sauerstoff-Gehalt der Umgebungsluft. Die Simulationen zeigen, dass die Leistungsfähigkeit und die Zyklenzahlen erhöht werden können, wenn Zink-Luftsauerstoff-Batterien mit reinem Sauerstoff und geringen Kohlendioxid-Verunreinigungen betrieben werden. Darüber hinaus sollte, je nach Anwendungsbereich und Umgebungsluftfeuchte, entweder die Luftfeuchte beeinflußt oder der verwendete Elektrolyt angepasst werden, um ein Austrocknen oder Fluten der Batterie zu vermeiden. In beiden Fällen müssten allerdings weitere Komponenten, wie zum Beispiel Pumpen, Reglereinheiten oder Filtereinheiten, verwendet werden, die den Vorteil der hohen Energiedichte dieses Batterietyps wieder verringern.

Die in dieser Arbeit eingesetzte Kombination von experimenteller und modellbasierter Analyse, liefert einen umfangreichen Einblick in die verschiedenen Reaktions- und Transportvorgänge in Zink-Luftsauerstoff-Batterien und ermöglicht ein besseres Verständnis der Funktionsweise dieses Batterietyps. Die eingesetzten Methoden sind erweiterbar und können auch für die Untersuchung weiterer Metall-Luft-Batterien eingesetzt werden.

Dem Anwenden muss das Erkennen vorausgehen.

- Max Planck -

1. Introduction

The limited amount of available conventional energy resources and the increasing energy demand in the 21st century require sustainable and efficient energy conversion. As a consequence thereof, further well-engineered energy storage devices for mobile and portable consumer electronics, automotive applications, and large scale energy storage are needed in the nearby future. This holds especially if energy is utilized from intermittent sources, such as wind power or solar power. Within this scope, battery technology is one promising field of energy storage. All aforementioned requirements might be fulfilled with improved batteries and battery systems in the upcoming years. The electrically rechargeable zinc air battery is, amongst others, one next-generation battery type that is able to cope with the future requirements of energy storage.

In the first chapter of this thesis, the working principle and the basic components of zinc air batteries are explained. In addition, the current state of the art of zinc air battery research is introduced. This chapter aims to elucidate the drawbacks and challenges for zinc air batteries. On this basis, the motivation for this thesis is pointed out and stated later on in chapter 2.

1.1. Composition and Working Principle of Zinc Air Batteries

In the zinc air battery (ZAB) considered for this thesis, zinc is stored within the battery, and oxygen is taken from e.g. the surrounding air to facilitate the overall reaction $Zn + 1/2\,O_2 \rightleftharpoons ZnO$. A schematic drawing of an electrically rechargeable ZAB is shown in figure 1.1. In general, ZABs consist of zinc electrode, air electrode, porous separator, and liquid electrolyte, which is usually an aqueous potassium hydroxide (KOH) solution. All battery components are embedded into a housing. The zinc electrode is confined by the housing and the separator, which is placed on top of the zinc electrode. The housing at the air electrode is open to the surrounding, and is usually equipped with air holes so that oxygen, the main reactant at the air electrode, can enter.

The air electrode is usually a gas diffusion electrode (GDE). It consist of a gas diffusion layer (GDL) and a catalytically active layer (CL), which is applied on one side of the GDL and faces the separator. The CL usually contains an electrically conductive material, such as carbon powder, and a catalyst. The purpose of the GDL is to ensure a defined flux of oxygen to the reaction zone at the CL. Therefore, the GDL is highly porous. The pores are partly filled with gas and are partly wetted with electrolyte. Additionally, the air electrode might be equipped with highly porous and hydrophobic polymer membranes facing the surrounding. The membranes are commonly made of PTFE (Polytetrafluoroethylene), and are applied to prevent that electrolyte and water vapor can leave the battery on the one hand, and that the optimal amount of oxygen is supplied to the CL on the other hand (see [1], p. 13.5).

The stored chemical energy in the zinc particles is converted in ZABs directly to electrical energy without intermediate steps of thermal and mechanical energy conversion by electrochemically separating the overall reaction into two parts, an oxidation and a reduction reaction. The reactions that occur are shown in figure 1.1: Solid zinc particles (Zn) and hydroxide ions (OH^-) are electrochemically converted to zincate ions ($Zn(OH)_4^{2-}$), which then precipitate to solid zinc oxide (ZnO), water (H_2O), and OH^- at the zinc electrode. At the air electrode, oxygen (O_2) and H_2O are electrochemically converted to OH^-. All reactions can be reversed. At each electrode, a current collector is applied to electrically conduct the generated

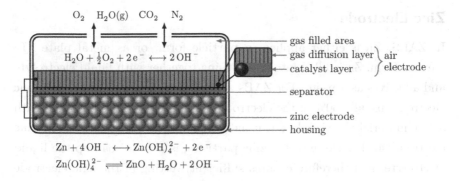

Figure 1.1.: Schematic drawing of an electrically rechargeable ZAB, showing the air electrode, the separator, the zinc electrode, the battery housing, the openings to the surrounding, and the reactions taking place. The liquid electrolyte (indicated with blue) is applied at the zinc electrode, the separator, and the air electrode.

and needed electrons for the electrochemical reactions. Both electrodes are electrically connected with each other through a load. Ions, OH^- in the case of ZABs, need to be transported between both electrodes through the separator to compensate the charge transport due to the transport of electrons. The same amount of H_2O is produced and needed at both electrodes, respectively. To maintain the reactions at both electrodes, it needs to be transported through the separator, or an excess of it, e.g. via the electrolyte, must be provided.

ZABs are operated at low temperatures around 298 K, and are either operated with gaseous oxygen or air. As illustrated in figure 1.1, ZABs can interact with the surrounding air. The surrounding air mainly contains nitrogen (N_2), carbon dioxide (CO_2), water vapor ($H_2O(g)$), as well as O_2, which can all enter or exit through the air electrode. More importantly, it can be seen that the same amounts of H_2O and OH^- are in each case provided and consumed at the respective electrode. Already at this point it becomes apparent that any loss or gain of the reactants and gaseous components due to the exchange with the surrounding air might influence the reactions and thus the operation behavior of ZABs [2].

Zinc Electrode

In ZABS, Zn is either applied in particle form, or as metal plate. In particle form, Zn is filled into the housing together with liquid electrolyte and additives as a paste. For ZABs equipped with such a paste-type zinc electrode, usually 50% of the electrode volume is void and not filled with active material [3]. Cellulose is usually included into the paste of the zinc electrode [4]. It surrounds the zinc particles, is able to soak up the liquid electrolyte, and therefore ensures sufficient wetting of the zinc electrode with electrolyte.

During battery discharge, the zinc particles and hydroxide ions are electrochemically converted to zincate ions in reaction (I) with the respective reaction rate r_I. The zincate ions can further precipitate to solid zinc oxide, water and hydroxide ions in reaction (II) with the respective reaction rate r_{II}. The reactions are as follows:

$$Zn + 4\,OH^- \overset{r_I}{\rightleftharpoons} Zn(OH)_4^{2-} + 2\,e^- \tag{I}$$

$$Zn(OH)_4^{2-} \overset{r_{II}}{\rightleftharpoons} ZnO\downarrow + 2\,OH^- + H_2O \tag{II}$$

As shown by Briggs et al., the primary zinc species in KOH-solution is the zincate ion, $Zn(OH)_4^{2-}$ [5]. According to the Pourbaix-diagram (potential-vs.-pH-diagram) for Zn (see [6]), Zn^{2+} ions are only predominant at a pH lower than 8 [7]. In the case of supersaturation of the electrolyte with $Zn(OH)_4^{2-}$, solid ZnO precipitates [8].

At elevated pH-values, $Zn(OH)_4^{2-}$ is predominant and solely the reactions (I) and (II) should occur. However, Zn is only electrochemically stable below -0.98 V vs. SHE (standard hydrogen potential), and thus zinc corrosion will occur when in contact with aqueous solutions such as KOH-electrolyte [7]. This can lead to an unwanted side reaction, the hydrogen evolution reaction (HER), which can be described as follows:

$$Zn + 2\,H_2O \rightleftharpoons Zn(OH)_2 + H_2\uparrow \tag{III}$$

This reaction has two negative effects on ZAB operation. It consumes the active material Zn, and forms hydrogen gas, which will expand in the sealed zinc electrode and might cause structural changes for the entire battery, or even complete battery failure.

Volume Change in Zinc Electrode

ZnO possesses a molar volume that is 1.6 times greater than that of Zn [9]. This implies that ZnO possesses a volumetric mass density that is 27% lower than that of Zn. A rather important issue for ZAB operation is revealed when looking at this difference: The particles in the zinc electrode presumably undergo a significant volume expansion during discharge when Zn particles are totally converted according to reactions (I) and (II).

Air Electrode

At the air electrode, oxygen dissolves in the electrolyte and reacts with water electrochemically to hydroxide ions during discharge. This reaction can be reversed for ZAB charge. The reaction in aqueous electrolyte, with the respective reaction rate r_{IV}, is as follows:

$$\frac{1}{2}O_2 + H_2O + 2\,e^- \xrightleftharpoons{r_{IV}} 2\,OH^- \qquad \text{(IV)}$$

This reaction is the key to enable the electrical recharge of ZABs. Thereby, oxygen is either consumed in the oxygen reduction reaction (ORR) or produced in the oxygen evolution reaction (OER), respectively.

Reaction (IV) occurs either as direct four-electron transfer reaction, or as two-electron peroxide-pathway reaction with formation of a peroxide intermediate, HO_2^-, [10]. In this thesis, solely the direct pathway is considered. Kinoshita [11] gives a comprehensive overview on the properties and mechanisms of the ORR and the OER, as well as on how catalysts enhance the electrocatalytic activity for each reaction mechanism.

For the electrical recharge of ZABs, a catalyst with bifunctional character, which enables both ORR and OER, is needed. In addition, the catalyst should reduce the polarization losses of the air electrode during battery operation. Since perovskite-based catalysts possess both aforementioned properties, they are commonly applied in electrically rechargeable ZABs [12]. Joerissen [10] presents a selective review on materials and synthesis principles for bifunctional air electrodes, and gives a detailed discussion about catalysts for bifunctional air electrodes. Recent research on air electrodes for ZAB operation and the catalysts applied will be further discussed in subchapter 1.2.

Battery Separator and Electrolyte

Separator

The separator in batteries is a selective barrier for certain reactants and products of each electrode; for example, gaseous O_2 should not pass through the separator of ZABs, otherwise Zn might oxidize directly. More importantly, the separator needs to electrically insulate both electrodes from each other, and has the task to ensure sufficient transport of OH^- ions between both electrodes. Besides, soluble intermediate species of the electrochemical reactions at zinc and air electrode should not be able to pass through the separator, since they might cause side reactions or affect the electrode reactions. However, Krej et al. [13] report that $Zn(OH)_4^{2-}$ ions are able to pass from the zinc electrode to the air electrode through separators that are presently used in ZABs.

In general, separators are highly porous and very thin membranes. They are composed of a polymer-backbone and certain additives. Since, batteries become more and more elaborate, the separators available are very versatile. Arora and Zhang [14] provide a good overview on the latter, explain the separator requirements, elucidate various separator types and structures, and list suitable separators for batteries with zinc electrode.

Separators can be characterized by their porosity and their tortuosity. Porosity quantifies the volume fraction of the separator that is not filled with separator material. For separators, the porosity, ε, is commonly quantified as the ratio between the void volume, V_{void}, and the total volume of the separator polymer-structure, V_{membrane}, so that:

$$\varepsilon = \frac{V_{\text{void}}}{V_{\text{void}} + V_{\text{membrane}}} \tag{1.1}$$

Tortuosity accounts for the fact that the pores of the separator are irregular in shape and position. The more tortuous a material is, the longer is the pathway for a permeating species through it. This implies that the pathway through porous material deviates from the straight pathway through it, which is accounted for with the tortuosity. The tortuosity, τ, can be described as the ratio between the permeation length, $l_{\text{permeation}}$, and the separator thickness, δ_{sep}, according to [14], so that:

$$\tau = \frac{l_{\text{permeation}}}{\delta_{\text{sep}}} \tag{1.2}$$

Porosity and tortuosity can be related to each other with empirical correlations. The most common one is the Bruggeman correlation [14], which is:

$$\tau = \varepsilon^{1-m^*} \tag{1.3}$$

Arora and Zhang mention values for the empirical parameter m^* to be between 1.5 and 4.5 for different types of separators [14]. Equation (1.3) is a common way to link both values, and to apply them in a correction factor for the diffusion of a species through a highly porous structure (see [15], p. 191). Instead of applying the diffusion coefficient, D, for the diffusion through the bulk (e.g. of the electrolyte), an effective diffusion coefficient, D^{eff}, is defined so that a correction for tortuosity and porosity is accounted for, so that:

$$D^{\text{eff}} = \frac{\varepsilon}{\tau} \cdot D \tag{1.4}$$

m^* from equation (1.3) is chosen to be 1.5 in this thesis where appropriate, so that equation (1.4) becomes:

$$D^{\text{eff}} = \varepsilon^{1.5} \cdot D \qquad (1.5)$$

Electrolyte

The ions produced and consumed in the battery need to be transported between both electrodes, so that the overall reaction is ensured. The medium of choice for the ion transport is the electrolyte. The electrolyte can either be a solid, a gel-type polymer, or a liquid. In ZABs it is common to apply liquid electrolytes. The liquid electrolyte is composed on the one hand of a dissociated species called solute and on the other hand of a solvent for the solute species. In aqueous liquid electrolytes, water is the solvent applied. The solute species is usually a dissociated salt, which is able to conduct ionic current for the ion transport needed between both electrodes.

The electrolyte applied is usually KOH-solution. Besides, different electrolyte solutions, based on different salts and the solvent water, with various viscosities and pH have been proposed for the application in metal-air batteries in general [7]. Of these, KOH-solution has been widely used as the electrolyte in ZABs because of its superior ionic conductivity, comparably large oxygen diffusion coefficients, and low viscosity [16]. Ionic conductivity values of KOH-solutions are broadly available [17, 18, 19]. A maximal ionic conductivity of 620 mS $\cdot$ cm^{-1} is reported for 6 M KOH-solutions [17].

Moreover, KOH-solutions are preferred because the reaction product K_2CO_3, produced during an unwanted side reaction, the carbonation reaction, possesses a larger solubility in KOH-solutions than in NaOH-solutions [16]. Therefore, it is less likely that salt precipitates in the structures of the air electrode.

However, the application of KOH-electrolyte has some disadvantages, especially when compared to non-aqueous electrolytes: Water in aqueous electrolytes can evaporate so that the battery can eventually dry out. Additionally, the alkaline KOH-solution can form carbonate species while losing hydroxide ions, which are crucial for the ion transport and the reactions, when in contact with carbon dioxide from ambient air. These issues will be addressed in more detail in chapter 7 of this thesis.

Besides aqueous liquid electrolytes, a material group known as ionic liquids has been recently introduced into ZAB research, as discussed by Harting et al. [20]. Ionic liquids are non-aqueous, and commonly possess a melting temperature below 373 K. Thus, they are in liquid state for common operation conditions of ZABs. Since they conduct ionic current without being dissolved in water, ionic liquids are attractive to overcome the above mentioned drawbacks of aqueous electrolytes. However, ionic liquids possess a major drawback. This type of electrolyte possesses an approximately 500 times lower ionic conductivity than the 6 M KOH-solution (1.26 mS $\cdot$ cm^{-1} at 298 K, see [20]). Besides, the reactants water and hydroxide ions are needed in the aforementioned electrode reactions in ZABs, which are usually not included in ionic liquids so that other reaction mechanisms might occur.

Polarization Behavior

The standard electrode potential using a caustic electrolyte at the air electrode is 0.401 V and -1.266 V at the zinc electrode, so that an overall open circuit potential (OCP) of 1.667 V can be obtained for ZABs [8]. The overall cell potential, E^{cell}, of a ZAB depends on the standard electrode potential of each electrode, E^0, on the overpotential, η, due to polarization and concentration changes at the respective electrode, on the set current, I, and on the ohmic resistance of the entire battery, R^{cell}, so that:

$$E^{cell} = E^{0,air} - E^{0,zinc} + \eta^{air} - \eta^{zinc} - I \cdot R^{cell} \qquad (1.6)$$

whereas the product $I \cdot R^{cell}$ is commonly denoted as ohmic drop.

The cell potential of ZABs drops significantly if they are polarized, i.e. if current is applied. This is in particular due to the sluggish ORR, which causes elevated overpotentials at the air electrode [7]. The reason is that

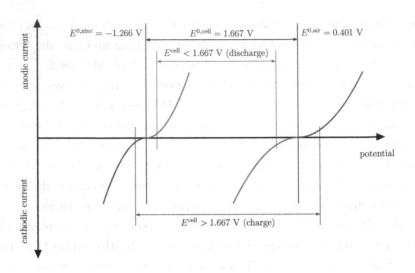

Figure 1.2.: Illustration of the cathodic and anodic current in ZABs as a
function of electrode potential; reproduced from Lee et al. [7].

the double bond of the oxygen molecule is remarkably strong and hard to
break, requiring catalysts to decrease the required activation energy [16].

Figure 1.2 illustrates the cathodic and anodic current as a function of
electrode potential for ZABs. During discharge, the available cell potential
is below 1.667 V; during charge, the cell potential must be above 1.667 V
to reverse the electrochemical reactions.

Commercially available primary ZAB button cells provide cell potentials
around 1 V during discharge (see [1], p. 13.7). They possess a flat cell
potential profile during discharge and can not withstand large discharge
current densities [16].

In theory, all aforementioned electrochemical reactions in ZABs are reversible. However, in practical state of the art ZAB applications the electrochemical reactions might not be reversible, and side reactions might
occur. As a consequence, the cycle numbers for charge and discharge of
ZABs in practical applications are currently limited. A brief history of the

development of ZABs and possible drawbacks of their application, as well as approaches to resolve them, will be discussed in more detail in the following subchapter.

1.2. State of the Art: Potentials and Drawbacks

The ZAB can be applied in a wide range of application areas, which is illustrated in figure 1.3. They might be applied either in portable applications for mobile electronic devices, in automotive applications to power cars or as range extender, and for stationary applications for example as energy storage to buffer the intermittent energy supply of wind or solar power plants. However, only if several drawbacks and challenges can be sustainably resolved, their application can be successfully realized.

ZABs appeal because they possess a multitude of advantages over existing battery systems. Since O_2 can be taken from the surrounding, ZABs offer a large theoretical energy density of 1353 Wh/kg (see [1], p. 1.13), which is approximately three times higher than for conventional lithium-ion batteries. Applying bifunctional catalysts, such as perovskite-based ones, at the air

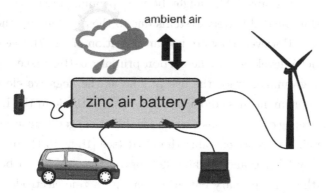

Figure 1.3.: Potential application fields for zinc air batteries: Portable applications, automotive applications, and stationary applications.

electrode can circumvent the use of expensive Pt-based catalysts. The applied active material Zn and the electrolytes applied are plentiful, and possess less safety issues than lithium-based batteries.

Currently, ZABs are commercially available as primary batteries for small consumer electronic devices, such as hearing aids. The successful commercialization of electrically rechargeable ZABs has failed up to this point due to two main reasons. On the one hand, the cycle stability is poor when compared to widely used 18650-type $C/LiMn_2O_4$ lithium-ion batteries; not more than 60 charge and discharge cycles without severe degradation are reported [21, 22] instead of 500 for $C/LiMn_2O_4$ batteries (see [1], p. 35.43). On the other hand, the half-open battery is by definition exposed to the surrounding air and is influenced by its content of relative humidity, carbon dioxide, and oxygen. This holds especially for electrically rechargeable batteries if they are meant to be operated as long as 500 charge and discharge cycles, and are exposed to the surrounding conditions over that period.

Paving the Way

Electrically rechargeable ZABs might be the upcoming alternative to state of the art lithium-based battery systems due to their remarkable theoretical energy density. However, they are not an invention of the 21st century. In 1866, Leclanché developed a zinc-carbon primary battery (see [1], p. 8.2) and laid the groundwork for the use of zinc as the negative electrode of batteries. Later, manganese dioxide, acting as a cathode in the Leclanché battery, was replaced by porous carbon, enabling oxygen transfer from the surrounding air and hence enabling the ORR (see [1], p. 13.1). In 1894, the U.S. patent by Walker and Wilkins [23] paved the way for carbon based GDEs as cathode for primary batteries that use a zinc electrode as anode. However, successful commercialization of primary zinc air batteries was hindered until the 1930s (see [1], p. 13.1). In 1932, Heise and Schumacher (see [24] and [25]) constructed a zinc air battery with alkaline electrolyte and a porous carbon air cathode, which was impregnated with wax to prevent flooding of the carbon pores (see [1], p. 13.1). Already at this point, the battery's sensitivity as result of the open air electrode became evident and attempts were made to circumvent this flaw with technical solutions. However, a systematic investigation on relative humidity, carbon dioxide and oxygen impact was not of primary concern.

During the 20th century, several technical improvements of the materials applied, as well as conceptual advances for ZABs were presented. Amongst these, the button cell configuration of ZABs [26], the zinc air slurry cell [27] and the efforts by the Israeli company Electric Fuel Ltd. on mechanically rechargeable ZABs for electromotive applications should be mentioned [8]. The design, development and commercialization strategies proposed by Electric Fuel Ltd. are well summarized [28]. Several reviews and book chapters on the technical progress and research interests on ZABs are available [8, 7, 29]. A multitude of recent review articles on the progress of ZAB research and technology underline that ZABs are one possible next-generation battery (see [16, 30, 31]). Moreover, high performance ZABs for automotive applications operated in fuel-cell mode, which function with continuous supply of air and a zinc slurry, are studied [32].

Drawbacks and Challenges

The research focus for ZABs is currently on electrodes and materials, as well as on understanding and manipulating the processes occurring within them. According to recent review articles by Caramia and Bozzini [30], and Pei et al. [31] on ZABs, the main drawbacks and challenges for electrically rechargeable ZABs are:

1. Efficient air electrode catalysts (see [30], p. 8)

2. Loss of carbon in the CL due to carbon corrosion, and hence performance issues for long-term operation (see [31], p. 21)

3. Shape change: Zinc electrode densification due to volume expansion of the oxidized zinc species (see [30], p. 8)

4. Shape change: Dendrite formation after a multitude of charge and discharge cycles (see [30], p. 8)

5. Loss of solvent of the aqueous electrolyte, respectively drying out of the battery due to water loss (see [31], p. 10)

6. Blocking of pores due to carbonate formation (see [31], p. 19)

Solely, Passaniti et al. name the flooding of the air electrode due to elevated relative humidity, the carbonation of the electrolyte, and the extradition to environmental conditions as *'major weakness of zinc air batteries'* (see [29], p. 13.17 and p. 13.31). They primarily discuss the factors that affect the service life of ZAB button cells. The reasons for the issues reported are not investigated in detail, and no counter-measures are proposed, for example by means of a model-based analysis for electrically rechargeable ZABs, so far.

Latest Materials Research

In the following, a survey on the most important research results on ZABs is given. Especially the year 2014 has brought up remarkable approaches and materials for the zinc electrode and the air electrode, which will be briefly summarized.

In general, the research on materials for ZABs has intensified during the past years. As a consequence, some of the aforementioned drawbacks have been successfully resolved, at least at lab-scale. Table 1.1 states selected operation properties for lab-scale ZABs that were reported between 2004 and 2014. It can be seen that ZABs are preferably operated at low current densities, and that only a low amount of the initial battery capacity can be withdrawn during discharge. This is primarily due to the aforementioned drawbacks of the materials at zinc and air electrode. In addition, the maximal discharge and charge cycle numbers that can be achieved until the cell performance decreases, were increased to promising values. This is primarily due to very recently reported improvements of air electrode and zinc electrode, which will be further elucidated in the following.

Table 1.1.: Selected discharge characteristics and cycle numbers for lab-scale zinc air batteries as reported in literature between 2004 and 2014.

reference and year	current density $[mA \cdot cm^{-2}]$	initial capacity $[mA \cdot h \cdot g^{-1}]$	maximal discharge capacity [%]	maximal discharge/charge cycles [-]
Yang et al. [33], 2004	-	360, 490	44.8, 60.5	-
Masri et al. [34], 2009	7.94	520	61.7	-
Drillet et al. [35], 2010	3.77	395	48.8	-
Masri et al. [36], 2013	7.94	556, 780	68.0, 95.0	-
Takeshita et al. [37], 2013	6	-	-	5-15
Ma et al. [38], 2014	15, 25	-	-	≈10
Lee et al. [21], 2014	-	-	-	60 (50 mA, 10 min per cycle)
Parker et al. [22], 2014	10, 24	640, 710	83.0, 89.0	45, zinc electrode only

Air Electrode

Since the reaction kinetics of the ORR at the air electrode are sluggish [16], the primary research is on catalysts for the air electrode. The aim thereby is to diminish the polarization losses during the ORR, while still enabling the OER with the same material, as well as ensuring the cycle stability of the supporting materials applied.

Several approaches utilize graphene oxide in combination with cobalt-oxalate-rods as bifunctional catalyst [39], iron nickelates on a carbon-fiber-matrix [40], carbon-free, Pt-free manganese-oxide structures [41], and per-ovskites, which are structurally manipulated to possess active oxygen-deficit sites [42]. The most promising results have been obtained by a Canadian research group (see [21] and [43]). They synthesized nanodisks of cobalt(II,III)-oxides and show its long-term stable application in ZABs. They report a cycle number of 60 without degradation of the catalyst. Accordingly, items 1 (efficient catalysts) and 2 (stable bifunctional catalysts) on the aforementioned list of drawbacks and challenges have been successfully addressed for air electrodes at lab-scale.

Zinc Electrode

Items 3 (zinc electrode densification) and 4 (dendrite formation) on the aforementioned list of drawbacks and challenges have been resolved on zinc electrode level. Namely by Parker and colleagues [22]; they also hold a patent on the same approach [44]. They report the first dendrite-free cycling of zinc electrodes, which they achieved by applying Zn-sponges (monolithic, three-dimensional and aperiodic in structure) as zinc electrode. They report 36 cycle numbers without dendrite-formation and significant shape change. However, they solely test their novel zinc electrode in a Ag-Zn cell and a Zn half cell. A full cell ZAB test is still to be shown.

To overcome the HER at the zinc electrode, the zinc particles are manufactured in a way that they possess a comparably large diameter (≈ 0.5 to 1 mm) and consequently a low surface area to suppress the HER. In addition, the surface of the zinc particles is usually doped with Bi, In or Pb. This increases the stability potential, and consequently decreases the HER (see [45]).

2. Motivation and Scope of this Thesis

The previous chapter has indicated that the electrically rechargeable zinc air battery has an enormous potential to be widely used for future energy storage. However, only limited charge and discharge cycle numbers can currently be achieved due to the unresolved issues mentioned. Almost all aforementioned issues of zinc air batteries that are addressed in research, are related to the materials applied or to the design chosen. Thus, they might be resolved with the help of material sciences and system design approaches. In this thesis, another approach is used: A detailed analysis of the reaction and transport processes that aims to gain a better insight into the operation of zinc air batteries and to understand how the cycle numbers can be improved. This is realized by means of a combined experimental and model-based analysis. The analysis helps to answer which battery composition and which air-composition should be adjusted to maintain stable and efficient charge and discharge cycles for zinc air batteries.

The thesis is structured into two main parts:

- Part 1 – Characterizing Reaction and Transport Processes

- Part 2 – Identifying Factors for Long-Term Stable Operation

In the first part of this thesis, electrochemical investigations and X-ray transmission tomography are applied on various zinc air battery set-ups. The experimental methods used are explained in more detail in chapter 3.

The physical and electrochemical processes in zinc air batteries are investigated for various battery and operation parameters to qualitatively analyze their impact on the diverse processes and interactions inside the battery. The experimental set-ups investigated and the measurements conducted are elucidated in chapter 4. The experiments reveal detailed information, e.g. on the cell potential and the solid and liquid species volumes in the zinc air battery, which is presented in chapter 5. Furthermore, a model-based analysis of the air electrode is presented in chapter 6 with the aim to qualitatively analyze the implications of the experimental findings obtained with electrochemical measurements and X-ray tomography. Thereby, the flooding of the air electrode with liquid electrolyte is investigated in detail to predict its impact on the overpotential, and the oxygen distribution within the air electrode.

In the second part of this thesis, the challenges that arise from the fact that zinc air batteries are half-open to the surrounding are addressed. As explained in subchapters 1.1 and 1.2, it should be of greater interest that ZAB operation is also affected by environmental conditions, such as relative humidity and carbon dioxide. This holds especially for electrically rechargeable batteries as they are operated over a long period with charge and discharge cycles, e.g. for the application in automotive applications where a long battery service-life is favorable. These issues are scarcely reported in present research, so that they first will be assessed theoretically in chapter 7.

Subsequently, the impact of the surrounding air-composition on electrically rechargeable zinc air batteries will be analyzed in this thesis by means of an expandable mathematical model, which is introduced in chapter 8. A scenario-based investigation of the operation of zinc air batteries is then conducted to identify to what extent the surrounding air-composition can influence the cycle performance of zinc air batteries. Thereby, the respective impact of relative humidity, carbon dioxide and oxygen content are the main factors of interest. The model-based analysis is intentionally not addressing the currently reported main challenges for the electrode materials in electrically rechargeable zinc air batteries. It is applied to elucidate the limitations of zinc air batteries that are exposed to the surrounding air.

The approach to reveal the expected operation limitations is as follows: First a basic model to account for the ideal operation of zinc air batteries is derived in subchapter 8.2. It is used to gain an in-depth understanding of the diverse processes and interactions, which are then compared to the experimental results obtained. Next, the basic model is modified to individually account for each of the air-composition impacts in subchapter 8.3, with the following scenarios:

- (a) Reference Scenario,

- (b) Relative Humidity Scenario,

- (c) Active Operation Scenario,

- (d) Carbon Dioxide Scenario,

- (e) Oxygen Scenario.

Each scenario is evaluated with transient simulations for constant current discharge and charge cycles, which is presented in chapter 9. The conducted simulations reveal detailed information, e.g. on the cell potential, the water and electrolyte content in each electrode, and the volumes of the solid and liquid species in the electrode of the zinc air battery. The underlying approach is versatile and expandable, includes only the aforementioned reaction and transport processes for zinc air battery operation and the air-composition impact on them. With this approach, a multitude of other electrochemical cells could be investigated as well.

Part 1 – Characterizing Reaction and Transport Processes

3. Basics of the Experimental Methods Applied

As mentioned in subchapter 1.2, a systematic investigation of the reaction and transport processes can help to gain a better understanding of ZABs, aiming to improve them further. In the first part of this thesis, experimental methods are chosen as tool to qualitatively assess the impact of various battery and operation parameters on the processes inside ZABs.

The following chapter introduces the fundamentals of the experimental methods used in this thesis. First, the basic electrochemical measurement techniques polarization curve, battery discharge, and electrochemical impedance spectroscopy are explained. Subsequently, the working principles and benefits of X-ray tomography are introduced to show why this technique is applied in this thesis to monitor the liquid and solid species in ZABs during operation. The practical application of these techniques for this thesis, with all parameters and experimental set-ups used, is elucidated in detail later on in chapter 4.

3.1. Electrochemical Methods

A multitude of methods can be applied to characterize electrochemical cells. State of the art methods are well described by Bard and Faulkner in their book *Electrochemical Methods - Fundamentals and Applications* [46]. Of these methods, polarization curves, battery discharge, and electrochemical impedance spectroscopy (EIS) are of particular interest to characterize the reaction and transport processes in ZABs for this thesis. These methods will be introduced in more detail in the following paragraphs.

Polarization Curve

Polarization curves are widely used to analyze electrochemical cells and their performance. They are usually applied to characterize fuel cells, but might also be used to investigate battery performance if the state-of-discharge (SOD) is not changed significantly during the measurement.

The measurement procedure for a polarization curve is as follows. Either the cell current or the cell potential is set constant, respectively. Then, the cell response to the set value is allowed to reach steady state, implying that each electrode is in electrochemical equilibrium. The respective other is allowed to reach a constant, or a (quasi)constant, value after a predefined time. Then the respective input value is increased or decreased after a fixed period of e.g. several seconds. Thereby, the respective response of the cell is measured. The input is increased or decreased stepwise and the procedure is continued until a predefined cut-off value is reached for the output value. Subsequently, each output value at the end of one step is allocated to the respective input value. To compare currents independently from cell size, usually the current density, i, is used for graphical depiction. Thereby, the applied or measured current, I, is divided by the cross-sectional area, A, of the electrochemical cell.

Figure 3.1 shows a typical polarization curve, i.e. cell potential as a function of current density, for an electrochemical cell with air electrode. In general, the polarization curve possesses three regions, labeled here as 1, 2, and 3. They are dominated by distinct effects on the electrochemical processes occurring (see [47], p. 143-145): Region 1 is predominantly determined by the activation of the electrochemical reactions at low current densities, which decreases the cell potential below OCP. This steep initial drop in cell potential is attributed to the initial hindrance for the electron and ion transfer occurring at the electrode/electrolyte interface at the elec-

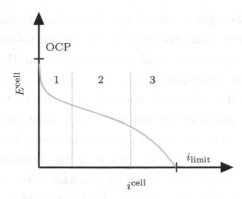

Figure 3.1.: Example of a polarization curve: Cell potential as a function of set current density: 1 - region of activation of the electrochemical reaction; 2 - region of linear decline of the cell potential; 3 - region of the reactants mass transport limitation with final cell potential decay.

trochemical double layer. With increasing current density in region 2, the cell potential is more and more dominated by the internal ohmic resistance of the electrochemical cell. There, the cell potential decreases almost linearly. In region 3, the mass transport of the reactants becomes limiting, and thus dominant at largest current densities. At these current densities, the reaction zone at the CL is only just sufficiently supplied with reactants, almost leading to a total decay of cell potential. The intersection of the polarization curve with the x-axis is where the cell potential approaches zero, which is usually labeled as limiting current density, indicated with i_{limit}. The calculation of i_{limit} for gas diffusion electrodes is given later on in chapter 7 in equation (7.7).

Battery Discharge and Charge

According to the *Handbook of Batteries* (see [1], p. 1.7), discharging a battery is defined as connecting the battery to an external load so that electrons are transferred from the battery anode, where the oxidation of one active material takes place, via the load to the battery cathode, where the electrons are accepted, and the reduction of the other active material takes place. In ZABs, a shortage of each reactant, needed in the electrochemical reaction described in subchapter 1.1, can terminate the discharge process. Most commonly, either the active material Zn or the reactant oxygen in the surrounding or supplied air are considered as limiting the discharge.

In this thesis, solely constant current discharge or charge is considered. To determine the withdrawn or input capacity, the constant current is multiplied by the duration for which the constant current was applied. To characterize a battery's state, commonly the state-of-discharge (SOD) is applied. The SOD is related to the withdrawn capacity, C_{Ah}, and the initially available capacity for battery discharge or charge, $C_{Ah,initial}$. It is calculated as ratio of the aforementioned capacities, so that:

$$\text{SOD} = \frac{C_{Ah}(t)}{C_{Ah,initial}} \cdot 100 = \frac{I^{cell} \cdot t}{m_{Zn} \cdot 819 \text{ mA} \cdot \text{h} \cdot \text{g}^{-1}} \cdot 100 \qquad (3.1)$$

where 819 mA $\cdot$ h $\cdot$ g^{-1} is the theoretical capacity for one gram of Zn (see [1], p. 1.11), and m_{Zn} is the mass of zinc particles that is initially in the zinc electrode.

Electrochemical Impedance Spectroscopy

Electrochemical impedance spectroscopy (EIS) is a standard technique to characterize processes occurring on different time scales within electrochemical cells. Amongst these processes, the short-term degradation, the calendric aging, the SOD impact, and the charge transfer, or diffusion processes can be investigated with EIS [48]. The use of equivalent circuits to quantify the results obtained with EIS is very common and well described [48].

For an EIS measurement, commonly a sine wave of either potential or current is chosen as excitation for the electrochemical cell of interest. If EIS is conducted galvanostatically, i.e. current controlled, the following procedure is applied (potentiostatic EIS is then analogous to the method mentioned before for the potential applied): During discharge or charge of the cell with a constant current density, i_{DC}, an alternating current density, $i_{AC}(t)$, with sine wave characteristics is superimposed [49], so that:

$$i_{AC}(t) = i_{AC}^{Amp} \cdot \sin(2\pi \cdot f \cdot t) \tag{3.2}$$

The cell potential of the electrochemical cell will then give a response to the input current density, which is likewise a sine wave. In this thesis, the amplitude of the input sine wave, i_{AC}^{Amp}, possesses a small enough value so that no higher harmonic responses should emerge in the response of the system, i.e. the system response to the sine wave is considered linear. Moreover, the response can be shifted in amplitude (Amp) or phase angle, ϕ, [49], so that:

$$E_{AC}(t) = E_{AC}^{Amp} \cdot \sin(2\pi \cdot f \cdot t + \phi) \tag{3.3}$$

The amplitude and the phase angle of the response depend on the chosen frequency f of the sine wave. The alternating cell potential $E_{AC}(t)$ is, as the aforementioned currents, superimposed to the direct E_{DC} signal measured.

The area specific impedance, Z, is then defined as ratio between potential and current density [48], so that:

$$Z(f) = \frac{E_{AC}^{Amp}}{i_{AC}^{Amp}} \cdot e^{j \cdot \phi} \tag{3.4}$$

Likewise, Z depends on the frequency of the sine wave applied and can be seen as a measure for the complex resistance of the electrochemical cell. Z is a complex number and can therefore be characterized by its modulus or magnitude $|Z|$ and its phase angle ϕ, or its real and imaginary part Z_{real} and Z_{imag}. It is to be noted that all impedance values mentioned in this thesis are related to the cross-sectional area of the battery electrode A, so that:

$$Z = \hat{Z} \cdot A \tag{3.5}$$

where $\hat{Z}$ is the measured impedance with the unit Ω, and Z is the area specific impedance in $\Omega \cdot cm^2$.

Commonly, an impedance spectroscopy is conducted by evaluating Z according to equation (3.4) for various frequencies. Typically f ranges from 1 mHz to 1 MHz for electrochemical applications [48].

To depict the analyzed impedance for the various frequencies, two diagrams are commonly chosen [49]: The Bode-plot and the Nyquist-plot. Both plots are equivalent and can be converted into each other. The Bode-plot illustrates the relation between $|Z|$ and f, both on logarithmic scale, as well as between ϕ and f, both on logarithmic scale as well. The Nyquist-plot shows the relation between Z_{real} and Z_{imag} for the frequencies recorded. In electrochemisty, the Nyquist-plot shows the negative imaginary part of Z on the y-axis in upwards direction [49].

A typical Nyquist-plot for an EIS of a zinc air battery stack is given by Ma et al. [38], and reproduced in figure 3.2. The closer a measurement point on the x-axis is to the origin of ordinates, the greater the frequencies are, at which the impedance was measured. In general, the following phenomena occurring inside the electrochemical cell can be revealed with the Nyquist-plot (see [49], p. 61): (a) Inductive behavior can be detected for measuring low values of Z_{real} and positive values of Z_{imag} at frequencies greater than 1000 Hz. Inductive behavior is attributed to the use of electrically conductive materials, either active material or current collector and most importantly to the wire-connections to the cell. (b) The ohmic resistance R^{cell} can be determined at the intersection of the curve with the x-axis; an ideal resistor possesses an imaginary part of Z equal to zero. R^{cell} includes the separator resistance and the electrolyte resistance, which can be correlated to the ionic conductivity of the electrolyte, and any ohmic resistance due to the presence of metallic parts in the cell, e.g. the current collector. (c) Capacitive behavior can be observed as loops or arcs in the Nyquist-plot. Capacitive effects can originate from a multitude of processes with different time scale, such as the charge of a double layer capacitance of an electrode, and the

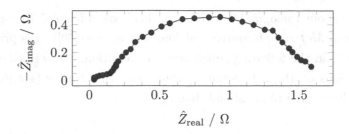

Figure 3.2.: Nyquist-plot for an electrochemical impedance spectroscopy of a zinc air battery stack. The general shape of the curve is given by Ma et al. [38] and reproduced with their data in this figure.

mass transport of ions in the electrolyte. These processes might also overlay and both contribute to the shape of an arc in the Nyquist-plot.

In ZABs, the arcs in the Nyquist-plot for low frequencies might also be influenced by processes with different time constants: The diffusion of oxygen through the GDL into the reaction zone at the CL, and the diffusion of OH^- in the liquid electrolyte through the porous zinc electrode. Thus, it is evident that EIS can not be used to clearly distinguish between processes with the same time constant. For this reason, an empirical method to analyze batteries with EIS is commonly applied: The state of the electrochemical cell is changed systematically by changing one parameter stepwise, e.g. the SOD or the cell temperature, whereas for each cell state an EIS is obtained. By means of the various EIS measured, the impact of the parameter varied can be interpreted by the change in the various EIS obtained. The realization of this procedure in this thesis will be described in subchapter 4.2.

3.2. X-ray Tomography

The following subchapter will provide the basics needed to understand the X-ray analysis applied in this thesis. In addition, materials that can be used for the set-ups for the tomography of electrochemical cells, and ZABs in particular, are discussed. Since Banhart gives a very comprehensive overview about tomographic methods in his book *Advanced Tomographic Methods in Materials Research and Engineering* (see [50]), it is primarily referred to in the following subchapter. In addition, Sprawls systematic explanations on the *Interaction of Radiation with Matter* (see [51]) are frequently referred to in this subchapter.

Imaging

Images are widely used to display detailed information of complex real objects in a comprehensible way. One advanced method of taking images is the tomography. It is a combination of taking two-dimensional radiographic projections of the object and subsequent image processing, which yields a three-dimensional image of the object investigated. Radigraphic images can be taken by using either electromagnetic radiation with the spectrum from microwaves to γ-rays, or particles such as protons, ions, electrons, and neutrons, or other waves. Therefore, tomography methods are commonly classified by the radiation applied.

The typical length scale for imaging methods comprises nine orders of magnitude of resolution and sample size [50]. Tomography can be applied in the 1-nm-range in the case of atom probe measurements, in the range up to 1 μm for electron tomography and X-ray 3D microscopy, in the range from 1 μm to 1 mm for X-ray μ-tomography, and on scales above 1 mm for neutron tomography (see [50], p. 16-17). In these ranges a huge variety of electrochemical systems and processes occurring within them can be visualized. Examples in the range of μm are the investigations of the local catalyst distribution in direct methanol fuel cells (see [52]) and the visualization of the water transport in the pores of a GDL (see [53]).

X-ray Transmission Tomography

Transmission tomography is well suited to analyze electrochemical cells since this technique is non-destructive and allows for an in operando analysis. However, certain difficulties are apparent, e.g. that special cell designs, such as the one presented in subchapter 4.1.1, have to be constructed if certain species inside the cell are of interest.

For X-ray transmission tomography, a beam of X-rays, a form of electromagnetic radiation, is sent out from a radiation source and directed to the object of interest. Banhart describes the properties and the procedure of acquiring an image with X-ray tomography as follows (see [50], p. 4-5): while passing through an object, the X-rays will interact with the object's atoms and then pass a detector system, which records the changes in X-ray intensity due to the attenuation of the X-rays after interacting with the object. Taking a single image with X-rays, an X-ray radiogram, yields a two-dimensional image of a cross-section of the object. Features within the object that are behind each other in beam direction are superimposed and are not distinguishable in one X-ray radiogram. To overcome this, more than one radiogram has to be acquired. In fact, the object is rotated around one axis and radiograms are recorded every given angular increment (often every degree over 360 degrees). The time needed to acquire one complete tomographic image can be in the range of hours, depending on the X-ray source parameters adjusted for the measurement.

In order to obtain a three-dimensional image from the radiograms recorded for every given angular increment, a computational procedure has to be applied. Mathematically, the obtained radiograms through a cross-section of the object can be seen as line integrals or pixel lines (see [50], p. 20). They are used to reconstruct a representative function of the cross-section of the object of interest, containing beam attenuation depicted as gray values. However, they might be noisy and incomplete due to streak artifacts, which increase the effort for reconstruction. Banhart gives an overview on the mathematics behind the reconstruction, as well as on correction methods for incomplete data (see [50], p. 23-29).

To interpret the reaction and transport processes within an electrochemical cell, it is common to apply the following procedure: A tomographic image of the electrochemical cell is taken at a chosen initial state, then the state of the cell is changed electrochemically, e.g. by discharging the cell, then another tomographic image of the cell is taken for the new state. This procedure allows to analyze the electrochemical cell more or less in operando and in a non-destructive way. In the end, the reaction and transport processes within the cell can be interpreted with respect to the differences of the taken tomographic images.

The generation of X-rays for the use in tomography for electrochemical cells is not discussed in-depth in this thesis. However, it is to be noted that in general two ways of generating X-rays are applied: One being the ejection of core electrons from atoms, where other electrons fall into the holes created and subsequently X-ray photons are emitted; the other being the acceleration or deceleration of free charged particles (see [50], p. 114-115). The technical application of these two methods is explained in more detail by Banhart (see [50], p. 115-119).

X-ray Interaction with Matter

X-rays are photons. They are elementary particles that possess energy. The energy of the photons is usually expressed in Joule or electron-volt (eV). The photons exhibit particle and wave properties at the same time. Due to this nature, three types of interaction are possible when photons pass an object: (1) penetration without interaction, (2) interaction by being completely absorbed and depositing all its energy, and (3) interaction by being scattered from the original pathway and depositing parts of its energy [51].

For the tomography of electrochemical cells, X-ray energies of about 5 to 150 keV are primarily applied since the active materials of interest often contain or are metals (see [50], p. 119). In this range of X-ray energies, two interactions between X-ray photons and material will predominantly occur: The photoelectric interaction and the Compton interaction. A comprehensive explanation of these physical effects is given by Sprawls (see [51], p. 3).

No matter what interaction occurs, it weakens the intensity of a photon beam passing through a material, which in its entirety is called attenuation. The attenuation depends strongly on the initial energy of the photons as well as on the material through that the photons travel. The total distance that a photon travels depends on the initial energy of the photon and on the material's density through that it is traveling [51]. In gases for example, the possibility of photon interaction and attenuation is less likely than in denser liquid or solid species. The total attenuation of a photon beam due to the aforementioned effects of scattering and absorption is expressed in the attenuation coefficient, μ (see [50], p. 112).

Material Impact on Attenuation

For tomography applications, the path of a single photon and the possible interactions on its way through an object are not of primary interest; rather the overall rate of photon interaction with the object is of primary concern [51]. Considering a beam of photons that enters a material part of one unit thickness, some of the photons will be removed due to interaction, independent whether photoelectric interaction or Compton, and some will pass through the object, reaching the detector [51]. The fraction of photons interacting per one unit thickness of material is called the linear attenuation coefficient, μ, with the unit m^{-1} [51]. Its value is a function of the energy of the individual photon, the atomic number, and the density of the object's material [51].

Occasionally the usage of the mass attenuation coefficient, μ^m, is more convenient since it is normalized by the material's volumetric mass density, ρ, so that solids, liquids, and gases possess the same unit for the coefficient (see [50], p. 113). It follows that:

$$\mu^m = \frac{\mu}{\rho} \tag{3.6}$$

If the object is homogeneously composed of various elements, the total mass attenuation coefficient is the weighted sum of the mass attenuation coefficients of the j single elements (see [54], p. 5), and can be expressed as:

$$\mu^{\mathrm{m}} = \sum_{k=1}^{j} w_k \cdot \mu_k^{\mathrm{m}} \tag{3.7}$$

where w_k is the mass fraction of element k.

The relation between the attenuation and the intensity of the photon beam can be described with the Beer-Lambert law of attenuation (see [50], p. 112): Thereby a beam with the intensity $\overline{I}_0$, which is directed through a sample possessing the linear attenuation coefficient μ and the sample thickness δ, will be directed to the detector behind the homogenous sample with the intensity $\overline{I}$, so that:

$$\overline{I} = \overline{I}_0 \cdot \exp^{-\mu \cdot \delta} \tag{3.8}$$

The intensity is a measure for the flux of photons of the electromagnetic waves applied (energy particles per time per area). With equation (3.8), the sample thickness δ_{half}, which is required to reduce the beam intensity by half, can be calculated. It is proposed by Waseda (see [54], p. 15-16) as:

$$\delta_{\mathrm{half}} \simeq \frac{0.693}{\mu} \tag{3.9}$$

With the database provided by [55] for μ-values of different materials, the material's density, and equations (3.7) and (3.9), it is possible to estimate a suitable material thicknesses required for a tomography set-up for the investigation of ZABs. The respective calculation is given in subchapter B.1 in the appendix. On this basis, δ_{half} is calculated for typical materials in ZABs at an X-ray energy of 50 keV. Zinc (Zn), zinc oxide (ZnO), 6 M KOH-solution, and a mixture of zinc and 6 M KOH-solution with a 1:1 ratio (Zn/KOH) are common materials inside the ZAB. The materials aluminum (Al), PTFE and PPS (Polyphenylene sulfide) can be applied as housing

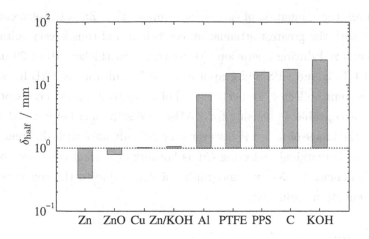

Figure 3.3.: Sample thickness needed to attenuate the intensity of a monochromatic X-ray beam by half at an energy of 50 keV calculated with equation (3.9).

for ZABs. Copper (Cu) might be applied as current collector in ZABs. The resulting δ_{half}-values for these materials are shown in figure 3.3. It can be observed that materials, which contain large amounts of carbon (the polymers PTFE and PPS, and pure carbon), and the KOH-solution yield highest values of δ_{half} due to their low attenuation. Next, samples of Al would yield a δ_{half}-value of half an order of magnitude lower than the aforementioned. Cu, the mixture of Zn/KOH, and ZnO exhibit δ_{half}-values of approximately one order of magnitude lower than for Al. Elemental Zn has the greatest atomic number within the chosen selection and thus the greatest attenuation at this X-ray energy. Consequently, δ_{half} for Zn is the lowest at about 0.4 mm. These results imply that Zn, ZnO, and the mixture of Zn/KOH yield on the one hand very reasonable attenuation difference in comparison to the other materials investigated but on the other hand would limit the sample thickness to avoid complete weakening of the X-ray beam.

All in all, the evaluation of δ_{half}-values implies that Zn is the element in the ZAB with the greatest attenuation coefficient and thus is very suitable to be visualized during operation. Moreover, even thicker (10 to 20 mm) layers of PTFE and PPS and graphite would be suitable as ZAB housing and as current collector, respectively. Thin (up to 2.5 mm) layers of Al could be acceptable as housing for ZABs; to distinguish between Al and PTFE in the image obtained is however very difficult since their difference in attenuation is marginal. Selecting Cu as housing or current collector would not be beneficial for X-ray tomography of ZABs due to the comparably large attenuation coefficient.

Energy Impact on Attenuation

The shown trend for δ_{half}-values of various materials is only valid at a designated energy of the X-ray beam applied. The implications for the tomography set-up design also depend on the X-ray source, respectively the X-ray or photon beam energy, used. In general, it can be observed that with increasing X-ray beam energy one of the aforementioned interactions on atomic level becomes predominant and is the main reason for the attenuation of the photons. In other words, the attenuation coefficients introduced will become energy-dependent (see [50], p. 114), so that two elements can possess the same attenuation behavior at the same X-ray energy. Figure 3.4 illustrates this issue. Thereby, the mass attenuation coefficient is depicted as a function of X-ray energy for the elements carbon (C), zinc (Zn), aluminum (Al), and the compound PTFE. The gray shaded area depicts the range of X-ray energies that are generated by the X-ray source applied in this thesis (see subchapter 4.2.3).

On the one hand, it can be observed that at 50 keV Zn possesses a greater mass attenuation coefficient than Al, which is beneficial for the X-ray tomography of commercial ZAB button cells because they possess an aluminum housing so that zinc particles can still be distinguished in the images obtained. On the other hand, it is evident that at 10 keV and

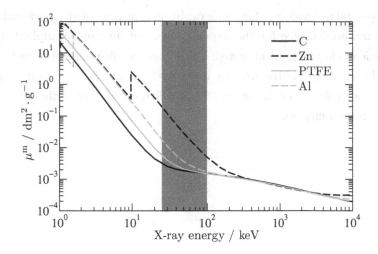

Figure 3.4.: Mass attenuation coefficient for various materials in ZABs as a function of X-ray energy; data taken from [55]. The gray rectangle indicates the energy range for the X-ray beam applied in the experimental part of this thesis in subchapter 4.2.3.

above 300 keV the difference in mass attenuation coefficients of Zn and Al is marginal, which is not favorable for X-ray tomography of ZABs because both materials might not be distinguishable.

In summary, tomography is a suitable tool to visualize the reactants and components in ZABs, and thus to analyze the reaction and transport processes within ZABs. For successful application of this method, it is essential to choose the X-ray energy and the contrast of attenuation coefficients of the materials applied in the tomography set-up as suitable as possible to the scientific question, e.g. the in operando observation of the active material of interest. In the case of ZABs, tomography might be applied to analyze the changes of the active material Zn during operation, and to monitor the liquid species inside the battery. In addition, tomography can reveal further information about the reaction of a single zinc particle.

The general methodology for tomography and the presented considerations on the material choice for the experimental set-up can be applied to a multitude of electrochemical systems. For example, lithium-sulfur batteries might be suitable for the investigation with X-rays because sulfur, the organic electrolyte containing carbon and lithium possess different mass attenuation coefficients.

4. Experimental Set-Ups and Measurement Details

In the previous chapter, the basics about the experimental methods applied in this thesis were given. In the following, the actual realization of these techniques for this thesis, with all parameters and experimental set-ups used, is introduced. The experimental analysis in this thesis is conducted to reveal detailed information e.g. on the cell potential and the solid and liquid species volumes in ZABs. However, not every experimental set-up is suitable for every measurement technique introduced. For this reason, in-house and commercial battery set-ups were selectively investigated. The different set-ups allow to gain detailed information on full cell, zinc electrode and air electrode level. Thus, the purpose of each set-up used in combination with a designated measurement technique, will be outlined as well in the following.

4.1. Set-Ups Applied

First, the set-ups, their design and the necessary electrode preparations for the zinc air batteries analyzed in this thesis will be explained.

4.1.1. In-House Zinc Air Batteries

Two in-house zinc air batteries were designed with the aim to adjust specific capacities and electrode components for each measurement. The electrode materials and electrolyte composition in these set-ups can be changed comfortably.

Set-Up for Electrochemical Characterization

One in-house battery set-up was designed (see also [56] and [57]) to electrochemically characterize batteries with various compositions of electrolyte, various zinc amounts (and thus capacities), and different catalyst layer, respectively gas diffusion electrode. In addition, this battery set-up can

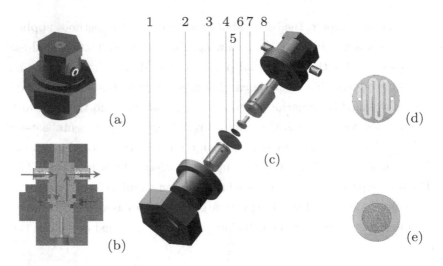

Figure 4.1.: CAD drawings of the zinc air battery set-up for the electrochemical analysis with (a) assembled battery set-up; (b) center plane cross-sectional view of the assembled battery with gas flow indicated; (c) exploded view of the battery with: 1 - union nut, 2 - PVC housing for zinc electrode current collector, 3 - zinc electrode current collector, 4 - separator, 5 - air electrode (GDL with CL), 6 - flow field at air electrode, 7 - air electrode current collector, 8 - PVC housing for air electrode current collector with gas flow inlet and outlet; (d) top view of the flow field applied at the air electrode; (e) top view of the zinc electrode current collector, the cavity is filled with a paste composed of electrolyte, active material, and additives. Reprinted from [57] with kind permission from Springer Science and Business Media.

be operated with active supply of air or pure oxygen. CAD drawings of the set-up are depicted in figure 4.1. The set-up consists of a two-piece housing, made of polyvinyl chloride (PVC), a PVC union nut, two gold-plated copper parts as current collectors, and a separator between the zinc and air electrode. The zinc electrode current collector was provided with a 1.3 cm diameter cylindrical indentation. The indentation in the zinc electrode current collector is adjustable in depth. A paste, containing the active material Zn, was filled into the cavity given by the indentation in the zinc electrode current collector. The paste consisted of Zn and ZnO (Grillo, Germany), Bi and cellulose-fibers (Sigma Aldrich, USA, German vendor), and liquid electrolyte. The solid constituents of the mixture were chosen in the weight-ratio of 10:1:1:1 (Zn:ZnO:Bi:cellulose). The weight-ratio of solid and liquid in the paste was chosen as 1:1 [57].

Batteries were prepared by filling the aforementioned paste in the cavity of the cylindrical indentation in the zinc electrode and assembling all components of the set-up. The theoretical capacities of the batteries prepared were approximately 300, or 400 mA $\cdot$ h for the indentation depths of 0.15 and 0.17 cm, respectively. The battery separator applied, Celgard 3401, was soaked with the respective electrolyte before assembly.

The air electrode was either a 0.2 cm thick commercially available GDL with Pt/C catalyst (W19216, Johnson Matthey, USA), or an in-house GDL (as described later on in this subchapter). The air electrode was placed on top of the air electrode current collector, which was provided with a 0.5 mm wide single-flow-channel for the supply of air or pure oxygen. Oil-free air was supplied with a mass flow controller (Smart Trak C100L, Sierra Instruments, USA) with a constant flow rate of 2.00 ml $\cdot$ min^{-1}.

Set-Up for X-ray Tomography

Gold-plated copper was used as current collector in the set-up shown in figure 4.1. Gold and copper are impractical for X-ray tomography because they possess greater attenuation coefficients for X-ray beams than the materials of interest inside the ZAB [58]. A detailed explanation about the choice of materials for the set-up is given in the appendix in subchapter B.1. On this basis, a special in-house battery was designed (see also [59]) to be analyzed with X-ray tomography. A CAD drawing of the set-up is depicted in figure 4.2. PTFE was chosen as housing for the battery electrodes. The housing consisted of two parts with an outer diameter of 1.5 cm, and a sealing cap at the air electrode with two air holes of 0.2 cm diameter and one hole for electrical connections. The zinc electrode current collector consisted of a cylinder-shaped 0.1 cm deep indentation of 0.6 cm diameter in a graphite/PPS case that was embedded into the PTFE housing. A paste, containing the active material zinc, was filled into the graphite/PPS case. A GDL, coated with a perovskite catalyst on one side, was used as the air electrode. Additionally, a stainless steel spring was included to ensure sufficient electrical contact between the electrochemical measurement device and the air electrode, as well as to ensure mechanical compression of the separator and the air electrode [59].

The air electrode had a diameter of 0.8 cm and a thickness of approximately 250 μm. It was prepared as suggested by Müller et al. [12]: A paste with isopropyl alcohol, VulcanXC72R (Cabot, USA) and $La_{0.6}Ca_{0.4}CoO_3$-catalyst (EMPA, Switzerland) was pasted on one side of a GDL (H2315, Freudenberg, Germany), dried at 333 K, and then coated with Nafion polymer solution (NS-5, Quintech, Germany). The catalyst loading adjusted was 1 mg $\cdot$ cm^{-2}.

The paste for the zinc electrode was prepared as follows. 207.1 mg Zn (Grillo, Germany) were mixed with 12.7 mg CM-cellulose (Sigma Aldrich, USA, German vendor). Subsequently, 87.4 mg of the dry mixture were filled into the graphite/PPS case. Then, 39.3 mg of a 6 M KOH-solution (Sigma Aldrich, USA, German vendor) containing 4 wt% ZnO (Grillo, Germany) were added. The applied battery separator, Celgard 3401 (Celgard LLC, USA), was soaked with the 6 M KOH-solution before assembly. The theoretical initial capacity of this battery was approximately 68 mA $\cdot$ h [59], which is based on the amount of Zn in the dry mixture filled in.

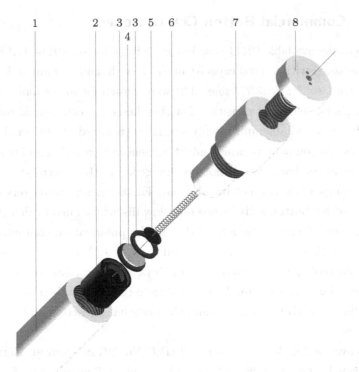

Figure 4.2.: CAD drawing of the zinc air battery set-up developed for X-ray
tomography; exploded view with: 1 - PTFE housing for the zinc
electrode, 2 - graphite current collector at the zinc electrode
where the zinc paste is filled in, 3 - O-ring sealing, 4 - separator
soaked in electrolyte, 5 - in-house air electrode (GDL with CL),
6 - stainless steel spring to establish electrical contact to the
air electrode, 7 - PTFE housing for the air electrode, 8 - PTFE
sealing cap with two air holes of 0.2 cm diameter and one hole of
0.4 cm diameter for electrical wiring. Reprinted from [59] with
permission from Elsevier.

4.1.2. Commercial Button Cell Batteries

Commercially available PR44-type button cells (IEC No. PR44, GP Batteries, Hongkong) with a rated capacity of 675 mA · h, and a nominal diameter of 1.14 cm (see [1], p. 13.7, table 13.2) were chosen for measurements with various discharge current densities. For this thesis, the commercial batteries were chosen as reliable and readily available standard. However, battery materials and compounds, which might influence the reaction and transport processes in the battery, can not be changed, e.g. the electrolyte applied and its composition can not be adjusted. Furthermore, the battery capacity is fixed for button cells. Since only low discharge current densities of 1.95 mA · cm^{-2} are recommended due to the polarization characteristics of button cells with 675 mA · h (see subchapter 1.1), complete discharge at this current density would take 14 days. This implies too long measurement times, if one intends to investigate the multitude of parameters that influence ZAB performance and the underlying reaction and transport processes.

Moreover, a ZA13-type button cell (IEC No. PR48, Conrad, Germany) with a rated capacity of 260 mA · h, and a nominal diameter of 0.77 cm (see [1], p. 13.7, table 13.2) was investigated with X-ray tomography by researchers of the Helmholtz-Zentrum Berlin [58], and evaluated in a joint publication [60]. The ZA13-type button cell is small in dimensions and possesses a thin housing of approximately 0.01 cm thick aluminum. Consequently, X-ray beams are not attenuated too intensively by the housing (see figure 3.3 in subchapter 3.2), so that X-ray tomography can be applied on this kind of batteries to visualize the active material zinc. However, not all materials within the button cells are stated by the manufacturer, which can lead to difficulties with the interpretation of the images obtained. In particular, the air electrode is usually contacted with a metal mesh, e.g. made of nickel, which possesses greater attenuation coefficients than the materials of interest inside the air electrode. By implication, the transport paths for oxygen and the liquid electrolyte at the air electrode could not be monitored during the discharge of this button cell type.

4.2. Measurement Details

Subsequently, the specifications of the electrochemical characterization and the X-ray tomography analysis applied on the previously introduced battery set-ups will be stated. Table 4.1 provides an overview about the electrochemical and X-ray measurements that were chosen for the different battery set-ups of various composition in this thesis. A detailed description of the parameters adjusted for each measurement method and the battery compositions used, is given subsequently in the respective subsections of this subchapter.

4.2.1. Electrochemical Characterization of In-House Batteries

All measurements for the in-house battery set-up in figure 4.1 were conducted at room temperature, which was approximately constant at 298 K, with a Reference 3000 galvanostat/potentiostat (Gamry, USA). Each in-house battery was subjected to a initialization procedure prior to the actual measurement as follows: Each prepared battery was subjected to current densities in steps of $1.88 \text{ mA} \cdot \text{cm}^{-2}$, with each current density being held for a period of 10 seconds, until the current density of $30.13 \text{ mA} \cdot \text{cm}^{-2}$ was reached.

In general, all EIS measurements for the in-house batteries were conducted at $0.75 \text{ mA} \cdot \text{cm}^{-2}$ constant current density and a $0.08 \text{ mA} \cdot \text{cm}^{-2}$ amplitude of alternating current for frequencies of $f = 0.01 - 10000$ Hz for 10 points/decade (see also [56] and [61]).

Table 4.1.: Overview of the electrochemical and X-ray measurements that were applied on the different battery set-ups in this thesis.

set-up	polarization curve	electrochemical impedance spectroscopy	constant current discharge	X-ray tomography
in-house set-up in figure 4.1	for various electrolytes, two different catalysts at the air electrode; subchapter 4.2.1	for various electrolytes, two different catalysts at the air electrode, and various SODs; subchapter 4.2.1	for various electrolytes at one current density; subchapter 4.2.1	-
button cells, PR44-type	-	for various SODs; subchapter 4.2.2	in climate chamber for various current densities; subchapter 4.2.2	-
in-house set-up for X-ray tomography in figure 4.2	-	-	between X-ray tomography measurements	for air electrode analysis; subchapter 4.2.3
button cell, ZA13-type	-	-	between X-ray tomography measurements	for full cell and zinc electrode analysis; subchapter 4.2.3

Variation of Electrolyte Composition

Various electrolytes were applied in the in-house set-up shown in figure 4.1 to investigate their impact on the electrochemical performance of ZABs. The preparation of the electrolyte solutions is explained in more detail in the appendix in subchapter B.2. In particular, batteries with 1.00 M, 6.07 M, and 10.03 M KOH-solution as electrolyte, respectively, were prepared. Additionally, batteries with pure 6 M KOH-electrolyte, and with 16.70 mol%, 28.60 mol% and 50.00 mol% of intentionally added potassium carbonate, K_2CO_3 were prepared and electrochemically investigated to determine the impact of the electrolyte composition with carbonates on the polarization curves, the impedance responses and the discharge behavior of ZABs. The molar amount added is expressed as molar percentage of K_2CO_3, and is determined as given in equation (B.1) in the appendix. The indentation in the zinc electrode current collector was set to 0.15 cm for the measurements with various amounts of K_2CO_3 in the electrolyte, and to 0.17 cm for measurements with pure KOH-electrolyte, respectively.

The polarization curves were measured, by applying current densities with increasing steps of 1.88 mA $\cdot$ cm^{-2} held for a period of 30 seconds. The SOD was thereby decreased marginally so that less than 1% of the initial capacity was withdrawn during the entire procedure. Solely the cell potential recorded after every 30 seconds and the set current density were taken to depict the polarization curves presented later on in the results chapter. Each polarization curve was measured three times to get an estimate error for the cell potential monitored via averaging. Subsequent to the polarization curves, EIS was measured to obtain the frequency response of the prepared batteries. Directly after the EIS, the very same batteries were discharged at a constant current density of 3.76 mA $\cdot$ cm^{-2} to monitor their cell potential with increasing SOD. The cell potential was thereby recorded every 60 seconds.

Variation of Catalyst at the Air Electrode

Two in-house batteries with different GDE, one with perovskite as catalyst and one with Pt as catalyst, were subjected first to polarization curve measurements, and then to EIS measurements with the same parameters as stated earlier in this subchapter.

Variation of SOD

One prepared in-house battery was subjected to an EIS measurement with the previously mentioned parameters. Afterwards, a constant current density discharge with 3.76 mA $\cdot$ cm^{-2} was conducted until a SOD of 63% was reached. Subsequently, another EIS with the same parameters was obtained to see the impact of the SOD on the impedance response (see also [56] and [61]).

4.2.2. Electrochemical Characterization of Button Cell Batteries

To investigate the impact of the discharge current density on the cell potential of ZABs, button cells from GP of the same lot number were operated in a climate chamber (HPP 108, Memmert, Germany) at 298 K and a relative humidity (RH) of 0.5, and were discharged separately at a current density of 1.95 mA $\cdot$ cm^{-2} and 4.90 mA $\cdot$ cm^{-2}, and one button cell from GP was operated in a temperature chamber (SU-641, ESPEC, Japan) at 298 K, and discharged at a current density of 9.80 mA $\cdot$ cm^{-2}. Thereby, two different automated battery test devices were used, a Series 4000 (Maccor, USA) and a Model 4200 (Maccor, USA), respectively.

One button cell from GP was subjected to a series of three EIS measurements with the Reference 3000 galvanostat/potentiostat at a constant current density of 2.93 mA $\cdot$ cm^{-2} and 0.29 mA $\cdot$ cm^{-2} amplitude of alternating current for frequencies of $f = 0.01 - 10000$ Hz for 10 points/decade. Each EIS was conducted at a different SOD, namely 1%, 24% and 44% to estimate the impact of the SOD on the impedance response of ZABs. The SOD was adjusted consecutively for each SOD with a constant current density discharge with the Reference 3000 at 9.80 mA $\cdot$ cm^{-2}. The button cell was kept inside a climate chamber (PL-3KPH, ESPEC, Japan) at 298 K and RH = 0.5 during discharge, and for the EIS measurements outside the climate chamber at a room temperature of 298 K.

For the commercially available button cells characterized in this thesis, the mass of zinc particles is not given. Thus, the SOD of button cells is estimated in this thesis based on the rated capacity stated by the manufacturer.

4.2.3. X-ray Tomography Specifications

X-ray Tomography for In-House Battery

A conventional X-ray tube (located at, and operated by researchers of the Helmholtz-Zentrum Berlin [58, 59]), was used for computer tomographic measurements of the in-house battery set-up shown in figure 4.2. The accelerating voltage was tuned to 120 kV while the current at the tungsten anode was adjusted to 83 μA. A 0.05 cm thick copper filter was applied in order to diminish beam hardening. 900 radiograms were taken for a full tomography over 360°. Each angle step was exposed twice for 1.8 seconds in order to increase the signal-to-noise-ratio. The radiograms were recorded with a Hamamatsu flat panel detector (2316 times 2316 pixel resolution). Tomographic measurements with an approximate duration of 110 minutes were [59]. Afterwards, the tomographic images were reconstructed and processed by researchers of the Helmholtz-Zentrum Berlin (see [59, 58]), and additional images were processed for this thesis with the software *ImageJ* [62].

For the in-house battery set-up shown in figure 4.2, the following procedure was applied: First, a X-ray tomography was conducted, subsequently the battery was discharged for 8.25 hours, then another X-ray tomography was conducted, then discharge was continued until the battery cell potential reached 0.7 V, and finally another X-ray tomography was conducted. To avoid further discharge during the X-ray tomography, OCP was maintained (zero current condition) for the battery during each tomography measurement. Thereby, a holding time of 500 seconds at OCP was maintained prior each tomography measurement. The battery discharge for this procedure was conducted galvanostatically at a current density of 7.07 mA · cm^{-2} with the Reference 3000 galvanostat/potentiostat. During the procedure, the air electrode was passively operated with ambient air at 298 K and RH = 0.4.

X-ray Tomography for Button Cell Battery

X-ray tomography measurements for a ZA13-type button cell were conducted by researchers of the Helmholtz-Zentrum Berlin [58] and the results, their interpretation, as well as a model-based analysis, are published in a joint publication [60]. The button cell was thereby stepwise discharged for approximately 210 hours in rheostatic mode, which is almost equivalent to a constant current density discharge of 5.4 mA · cm^{-2} (see also [60]). During the measurement, the air electrode was passively operated with ambient air at 295 K.

The same aforementioned conventional X-ray tube was used for the to-mographic measurements with the ZA13-type button cell. The accelerating voltage was adjusted to 100 kV while the current at the tungsten anode was set to 100 μA [60]. The spectrum of X-ray energy was ranging from 25 keV to 100 keV, with the highest beam intensity at approximately 40 keV. A 0.1 cm thick copper filter was applied in order to diminish beam hardening. 1200 radiograms were taken for a full tomography over 360°. Each angle step was exposed for 2.2 seconds three times to increase the signal-to-noise-ratio. The radiograms were taken using the same flat panel detector as mentioned before. Tomographic measurements with an approximate duration of about 3.5 hours were taken at initial battery state, after several discharge steps (10 in total) and finally at the end-of-life of the battery. Subsequent to the measurements, the tomographic images were reconstructed and processed using the software *Octopus* by researchers of the Helmholtz-Zentrum Berlin (see [60, 58]), and additional images were processed for this thesis with the software *ImageJ* [62].

The discharge process was stopped during the tomographic measurements and continued afterwards. Tomographic measurements were started after switching off the current load (zero current condition) and after reaching a constant cell potential [60]. This was done to avoid particle movement and zinc conversion that would occur during continued discharge and would lead to uncorrectable errors in the reconstructed tomography [60].

5. Experimental Results and Discussion

In the following chapter the electrochemical measurement and X-ray analysis results are presented. These experimental results will reveal detailed information about the operation of ZABs, which will then be correlated to the electrochemical and chemical reactions, and the transport of reactants and electrolyte inside the battery.

Each battery set-up and each measurement method was selected specifically to demonstrate the impact of various battery properties and operation parameters on the processes inside ZABs. Thus, the results for the general reaction processes of ZABs are addressed first, i.e. processes that can be monitored with the cell potential and impedance response. Secondly, the results for the processes at the zinc electrode are presented, i.e. processes that can be monitored with X-ray tomography and that can be attributed to distinct frequencies of the impedance response. Thirdly, the results for the processes at the air electrode are presented, i.e. processes that can be visualized with X-ray tomography, and that can be indirectly shown with the analysis of cell potential and impedance response.

In general, all reaction and transport processes in ZABs are affected by the operating temperature. However, the temperature impact on ZABs is not in the main focus of this thesis. For the sake of completeness, polarization curve measurements and EIS responses of button cell ZABs were investigated at various temperatures. The respective results obtained are not included in this chapter, but are briefly presented in subchapter B.5 in the appendix.

5.1. Basic Processes

In the following, the results of polarization curve measurements, electrochemical impedance spectroscopy measurements, and discharge curves are described and discussed. They will reveal the overall cell response, i.e. expressed by means of cell potential and impedance. This analysis will help to qualitatively assess the basic processes that occur for ZAB operation, and how they are influenced by the electrolyte used and the current density applied. This subchapter is based on the publication [57] from which experts are used with kind permission from Springer Science and Business Media.

Polarization Behavior

For the results in figure 5.1 (a), the battery set-up was operated with 1.00 M, 6.07 M and 10.03 M KOH-electrolyte, respectively, to reveal the impact of the KOH-molarity on the polarization behavior of ZABs. The standard deviation for three consecutive electrochemical measurements is indicated as error bar. It can be observed that batteries with 6.07 M and 10.03 M possess about the same cell potential. The curve for 10.03 M is showing slightly lower cell potential at current densities above 15 mA $\cdot$ cm^{-2}. The measurement for 1.00 M electrolyte reveals a much steeper decline of cell potential with increasing current density than for the other measurements. This might be due to the comparably low concentration of OH$^-$. On the one hand, it can evoke a greater ohmic drop due to the lower ionic conductivity at this concentration, and on the other hand it can slow down the reactions at the zinc electrode, which causes a greater overpotential due to polarization. A distinct quantification of these two effects is given with the analysis of the electrochemical impedance spectroscopy measurements later on in this subchapter.

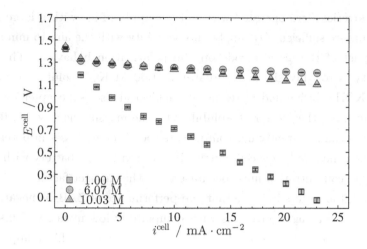

(a)

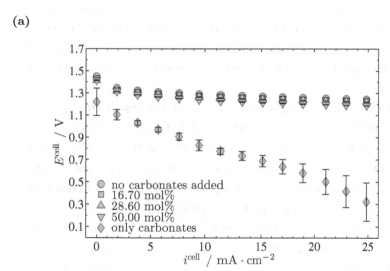

(b)

Figure 5.1.: Polarization curve measurement results for: (a) three different molarities of KOH; (b) mixtures of 6 M KOH-solution without, and with added K_2CO_3, indicated by the molar amount of added carbonates. SOD close to 100%. Reprinted from [57] with kind permission from Springer Science and Business Media.

All batteries were operated with active supply of air with a large flow rate to ensure sufficient O_2-supply and solubility with the aim to minimize the impact of the air electrode on the polarization behavior. The O_2-solubility is roughly six times greater in 1.00 M KOH-solution than in 6.07 M KOH-solution under presence of ambient air (see subchapter 7.3 and [63]). However, the greater O_2-solubility for the measurement with 1.00 M electrolyte can apparently not compensate the observed losses in potential. Thus, the comparably low cell potential measured for the battery with this electrolyte concentration must be caused by other impact factors.

Overall, the results indicate that the performance of ZABs is negatively affected when using electrolytes with comparably low molarity. This implies that the usage of elevated molarities is essential to achieve improved polarization performance. However, this result is preliminary and solely based on the polarization curve measurements presented; for the operation of electrically rechargeable ZABs additional factors, such as the impact of the surrounding on the open air electrode, need to be taken into account.

Figure 5.1 (b) shows polarization curves for in-house batteries operated with 6 M KOH-electrolyte without added K_2CO_3, 6 M KOH-electrolyte with added K_2CO_3 (16.70 mol%, 28.60 mol% and 50.00 mol%), and with pure K_2CO_3-solution. This analysis aims to reveal the impact of carbonates, which might form during long-term operation of electrically rechargeable ZABs, on the electrochemical performance of ZABs.

The polarization curves measured with 16.70 mol% up to 50.00 mol% of carbonates only marginally deviate from the results obtained for 6 M KOH-electrolyte without added carbonates, and from the result for pure 6.07 M KOH-electrolyte in figure 5.1 (a). This can be explained by the sufficiently large concentration of OH^-, which ensures that a sufficient amount OH^- is available for the reaction at both electrodes. In addition, the polarization curve is hardly affected by increased percentages of carbonates added in the quasi ohmic region at intermediate current densities. This can be explained for example by the fact that the ionic conductivity of the 6 M KOH-electrolyte with 50.00 mol% is only 33% lower than for 6 M KOH-electrolyte (see [57]), which would still lead to a 3-fold increase in ohmic drop.

Moreover, it is shown that a ZAB operation with pure K_2CO_3-electrolyte is not favorable since the cell potential is much lower for all measurement points, and the ohmic drop is much larger than for the other measurements. Besides, the measurement errors are larger than for the other batteries. These observations might be attributed to the fact that the electrochemical reactions (I) and (IV) for ZABs do not hold for pure K_2CO_3-electrolytes where only marginal amounts of OH^- are present as impurity or from the solvent water. By implication, different reactions and thus different cell potentials might occur for the operation of ZABs with pure K_2CO_3-electrolyte.

The ZAB performance shown with the presented polarization curve measurements in this subchapter suggests that K_2CO_3 has a weaker influence on the cell potential in ZABs than the amount of OH^- in the electrolyte. This implies that adding up to 50.00 mol% of carbonates to an electrolyte, while keeping elevated molar amounts of K^+, may still result in acceptable cell performance. Using this result as starting point, the idea of adding K_2CO_3 to the alkaline electrolyte of ZABs will be further investigated with the model-based analysis of electrically rechargeable ZABs later on in subchapter 9.2 to reveal the benefits for the cycling stability of ZABs that might arise from the use of K_2CO_3 in the electrolyte.

Electrochemical Impedance Spectra

The following EIS results are used to further elucidate the reason for the previously discussed polarization behavior trends, as well as to interpret the impact of the KOH-concentration and the added K_2CO_3 in the electrolyte on the reaction and transport processes at both electrodes that might appear in the impedance response.

Figure 5.2 (a) depicts the Nyquist-plots of the impedance response of ZABs prepared with 1.00 M, 6.07 M and 10.03 M KOH-solution. In all measurements, there is one pronounced arc in the low frequency region, and there is another smaller, less pronounced arc in the high frequency region (see enlargement). The curves strongly deviate from each other above $Z_{real} > 10 \ \Omega \cdot cm^2$, possessing a larger impedance for decreasing electrolyte concentrations of KOH. This is particularly distinct for the measurement result with 1.00 M KOH-solution.

The enlargement in figure 5.2 (a) shows the impedance response in the high frequency region. The area specific ohmic resistance at the highest frequency is 0.7481 $\Omega \cdot cm^2$ for a battery with 1.00 M KOH-solution, 0.3723 $\Omega \cdot cm^2$ for 6.07 M and 0.3844 $\Omega \cdot cm^2$ for 10.03 M. These values can be seen as internal battery resistances. The internal battery resistance is the highest at 1.00 M KOH-electrolyte and the lowest for 6.07 M KOH-electrolyte. This trend is in line with ionic conductivity data for KOH-solutions (see [17]), so that it might be presumed at this point that the internal battery resistance is impacted by the presence of OH^- in the electrolyte. It is to be noted that the impedance of porous electrodes behaves significantly different than for flat electrodes [64], so that at high frequencies, the external electrolyte resistance in the porous zinc electrode depends predominantly on the electrolyte applied [65]. To estimate the contribution of the porous zinc electrode in the battery to the overall internal battery resistance, equation (B.3) in the appendix is used. The calculation shows that 74% of the internal battery resistance for the EIS measurement with 6.07 M KOH-electrolyte originate from the electrolyte in the zinc electrode. Consequently, the two times lower area specific ohmic resistance for the measurements with 6.07 M KOH-solution compared to 1.00 M KOH-solution can mainly be attributed to the superior ionic conductivity of the 6.07 M KOH-solution. This is also reported for impedance measurements of zinc electrodes in KOH-electrolytes with various molarities [66].

Furthermore, the values of the internal battery resistances obtained imply that for example polarization losses at 25 mA $\cdot$ cm^{-2} are approximately between 9 mV and 18 mV for the KOH-molarities investigated. However, this is a marginal contribution to the overall polarization losses and might

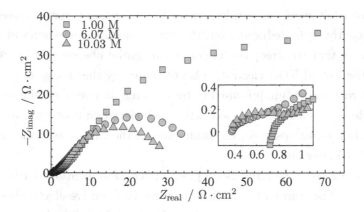

(a)

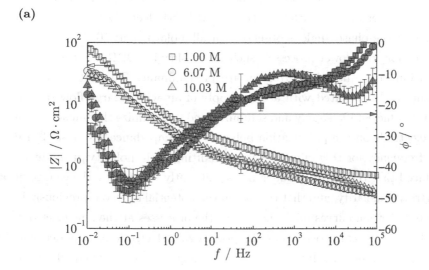

(b)

Figure 5.2.: EIS results for different molarities of the electrolyte: (a) Nyquist-plots for batteries with 1.00 M, 6.07 M and 10.03 M KOH-solution applied; (b) Bode-plots for the same batteries. Reprinted from [57] with kind permission from Springer Science and Business Media.

be neglected at every current density for the polarization curves presented. Consequently, it is deduced from the impedance measurements at high frequencies that the steep decline of cell potential observed in figure 5.2 (a) for the 1.00 M KOH-electrolyte is not caused by the change of internal battery resistances but presumably by a change in overpotential at zinc electrode or air electrode. To allocate which electrode is determining the polarization losses observed, the Bode-plots of the EIS measurements will be analyzed later on.

Figure 5.2 (b) shows the Bode-plots of the aforementioned EIS measurements. The standard deviation for three consecutive electrochemical measurements is indicated as error bar. The modulus of the impedance is similar for the entire range of frequencies for 6.07 M KOH and 10.03 M KOH, but is greater for the battery with 1.00 M KOH-electrolyte. A minimum value for the phase angle is observed for all molarities at 10^{-1} Hz. There, presumably the slower processes, such as the sluggish ORR and the oxygen diffusion in the liquid electrolyte, are predominant. The batteries were intentionally operated with active supply of air with a large flow rate to ensure sufficient O_2-supply and solubility, and to minimize the impact of the air electrode on the polarization behavior and impedance response. Since the frequency for this minimum is not shifting with molarity, it might be deduced that the air electrode is not significantly affected by the change of electrolyte molarity, and that the loss in cell potential observed previously in the polarization curves might be due to the processes at the zinc electrode.

Electrochemical processes at the porous zinc electrode usually occur at frequency larger than 10^2 HZ [67]. In the frequency region from 10^2 Hz to 10^4 Hz, it can be observed that the phase angle increases with increasing molarity of KOH. This is presumably attributed to the zinc electrode kinetics, since the reaction rate at the zinc electrode is enhanced at elevated molarities [68].

Figure 5.3 (a) depicts the Nyquist-plots for batteries prepared with electrolyte containing various added molar percentages of carbonates. All possess one pronounced arc in the low frequency region, and another smaller, less pronounced arc in the high frequency region (see enlargement). The arc at low frequencies first decreases and then increases its size with increasing amounts of K_2CO_3. Thus, adding K_2CO_3 to the electrolyte might enhance the kinetics of the oxygen reduction reaction for small portions added.

The enlargement in figure 5.3 (a) shows the impedance response in the high frequency region. The area specific ohmic resistance at the highest frequency is 0.1711 $\Omega \cdot cm^2$ for a battery with no added carbonates, 0.2223 $\Omega \cdot cm^2$ for 16.70 mol% added carbonates, 0.2557 $\Omega \cdot cm^2$ for 28.60 mol% added carbonates and 0.2781 $\Omega \cdot cm^2$ for 50.00 mol% added carbonates. As mentioned earlier for the EIS measurement with various KOH-electrolyte, the internal battery resistance is predominately determined by the ionic conductivity of the electrolyte molarity applied. In this case, this implies that more carbonates in the electrolyte yield a lower ionic conductivity and consequently a greater internal battery resistance.

It is to be noted that the ohmic resistances for the two measurements presented with pure 6 M KOH differ (0.1711 $\Omega \cdot cm^2$ and 0.3723 $\Omega \cdot cm^2$, respectively). This can be explained by the fact that the prepared zinc paste was different in amount and that the indentation depth in the current-collector of the zinc electrode was different (see subchapter 4.1.1), which consequently results in a different impedance response of the zinc electrode. The same holds for the observations in the Bode-plots presented.

Figure 5.3 (b) shows the Bode-plots of the aforementioned EIS measurements with added carbonates. The modulus of the impedance is almost identical for the entire range of frequencies for all electrolytes applied. A minimum value for the phase angle is observed for all molarities at 10^{-1} Hz (the same frequency as for the measurements with various KOH-electrolyte), which is where the slower processes at the air electrode are expected. For frequencies from 10^3 Hz to 10^4 HZ, it can be observed that the frequency at which a maximum value for the phase angle is emerging, is increasing slightly with increasing percentage of K_2CO_3 added. This is presumably due to the presence of an increased amount of CO_3^{2-} that possibly enhance

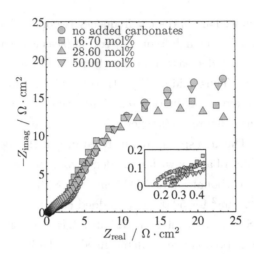

(a)

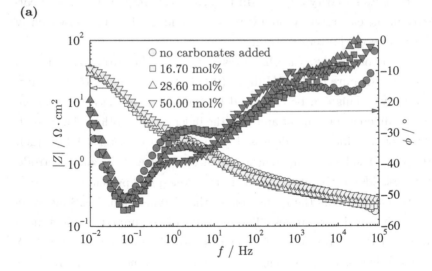

(b)

Figure 5.3.: EIS results for different percentage of added carbonates in the
6 M KOH-electrolyte: (a) Nyquist-plots; (b) Bode-plots for the
same batteries. Reprinted from [57] with kind permission from
Springer Science and Business Media.

the formation of $ZnOH_4^{2-}$ in reaction (I) and slightly improve the kinetics of the reactions at the zinc electrode. Additionally, the frequency at which a maximum value for the phase angle can be observed in the range of 10^0 Hz to 10^1 Hz, is decreasing with increasing percentage of added K_2CO_3. The frequency observed is the lowest for 50.00 mol% of K_2CO_3. This observation might imply that another charge transfer, and thus another reaction, slower than reaction (I), takes place if large amounts of K_2CO_3 are added to the electrolyte.

Compared to the impedance measured for various KOH-electrolytes, it can be concluded that the molar percentage of added K_2CO_3 affects the processes at the zinc electrode to a smaller amount than the concentration of OH^-. However, the added K_2CO_3 might affect the reactions at the zinc electrode significantly, which in turn could change the discharge performance of the entire battery.

Discharge Behavior

The aforementioned results revealed information about the short-term performance of ZABs, about the reaction kinetics, and about the transport processes in ZABs. The results could not reveal information about the discharge behavior of ZABs. For this purpose, selected batteries were discharged for various electrolyte compositions and various current densities.

First, electrolytes of 6.07 M KOH-solution, and of solutions with 6 M KOH and 16.70 mol% and 50.00 mol% of added K_2CO_3 are used to prepare batteries and discharge them at a constant current density of $3.76\,mA \cdot cm^{-2}$. Figure 5.4 illustrates the cell potentials obtained. The discharge curves for the battery with 6.07 M KOH-solution, and with 6 M KOH-solution and 16.70 mol% added K_2CO_3 possess a flat cell potential at approximately 1.28 V until a SOD of approximately 60% is reached. Then, the cell potential decays below 1.00 V, indicating the maximal SOD for this measurement. The maximal SOD for the battery with 6.07 M KOH is 70%, which is in line with other experimental findings for this current density [3, 29]. The

discharge for a battery with 6 M KOH-electrolyte with 16.70 mol% added
K_2CO_3 shows a maximal SOD of 62%.

On the contrary, the cell potential obtained for the battery with 6 M KOH
and added 50.00 mol% of K_2CO_3 behaves considerably different from the
case with pure KOH-electrolyte. After an initial drop to 1.23 V, the cell
potential recovers to approximately 1.28 V, and then decreases until the
cell potential is below 1.00 V at approximately 40% SOD. This leads to
the conclusion that 50.00 mol% of added K_2CO_3 might reduce the maximal
SOD achievable at this current density by half. This is presumably caused
by the participation of CO_3^{2-} in the electrode reactions, which was deduced
from the EIS measurements shown in figure 5.3 (b). To analyze the shape
of the discharge curve and the underlying changes in the reactions at both
electrodes is out of the scope of this thesis, and remains an open issue for
future work.

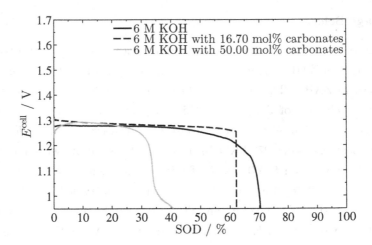

Figure 5.4.: Discharge curves for the in-house set-up for different elec-
trolyte compositions; constant current density discharge at
$3.76 \, \text{mA} \cdot \text{cm}^{-2}$. Reprinted from [57] with kind permission from
Springer Science and Business Media.

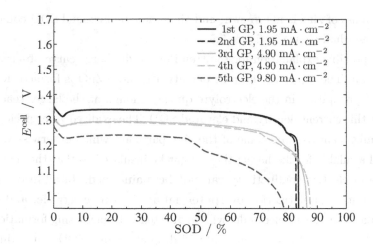

Figure 5.5.: Discharge curves for GP button cells for different current densities applied.

To obtain further results for various discharge current densities, commercially available button cells were chosen because they are a readily available and reliable alternative to the ZAB in-house set-up. The button cells were discharged at current densities of 1.95, 4.90, and 9.80 mA · cm^{-2} to resolve its impact on the maximal SOD achievable. The results obtained are shown in figure 5.5.

All discharge curves obtained possess a flat cell potential, as explained in subchapter 1.1. With increasing current density, the cell potential observed for example at 30% SOD decreases from 1.35 V to 1.29 V and to 1.21 V, respectively for each curve, which is in line with the polarization curve measurements shown in figure 5.1 (a). The maximal SOD obtained is the highest with 87% for the lowest current density applied, and is approximately 10% less for the highest current density of 9.80 mA · cm^{-2} applied. It is to be noted that the SOD for the commercial button cells is expressed with the rated capacity of 675 mA · h given by the manufacturer in this thesis. The actual mass of zinc is not known, and thus the true maximal SOD could be much less than shown in figure 5.5.

One possible explanation for the trend of the discharge curve observed is as follows. The larger the current density, the more ZnO is formed, which can not precipitate in the electrolyte quickly enough at the larger reaction rate at this current density and elevated SOD. The oxidized zinc species, i.e. ZnO, might encapsulate some of the zinc particles, which otherwise would be still available for discharge; they might be insulated so that the required ionic and electric conductivity can not be maintained. In addition, it is likely that the polarization overpotential at the zinc electrode, and also presumably at the air electrode, will be affected with increasing formation of ZnO. As a result, the cell potential will drop at smaller SOD values during the discharge with elevated current densities. A better understanding of this mechanism, and on how the reaction products will be deposited around the zinc particles is provided in subchapter 5.2.3 with the X-ray tomography results. In combination with the here presented electrochemical analysis, this investigation will help to further understand the discharge behavior of ZABs. All in all, the discharge profiles in figure 5.4 and 5.5 confirm that primary ZABs are currently only an option for low power applications.

5.2. Zinc Electrode Processes

In the following, the results obtained for experiments that were chosen to influence the processes at the zinc electrode are presented and discussed. The results will help to understand the changes inside the zinc electrode during discharge, and allow for a qualitative description of the volume expansion of the zinc electrode, and the consequences for the transport of the liquid electrolyte. First, the impedance response of ZABs at various SOD is presented. Second, X-ray tomography results for a single zinc particle that is converted with increasing SOD are presented. Third, X-ray tomography results of a button cell for varying SOD are elucidated.

5.2.1. Impact of State-of-Discharge

As shown with the discharge curves in subchapter 5.1, the cell potential of
ZABs can change significantly at greater SOD values, which is presumably
due the conversion of the active material zinc and the concurrent formation of
oxidized zinc species. To show the impact of the active material conversion on
the impedance response of ZABs, two batteries (one commercially available
button cell and the in-house set-up) were discharged stepwise, which should
evoke different electrochemical and chemical states of the electrodes, and
were monitored with EIS measurements at the SOD values adjusted.

Figure 5.6 (a) depicts the Nyquist-plots for the impedance response of the
prepared in-house battery, evaluated at a SOD of 1% and 63%, respectively.
It can be observed that the size of the arc in the low frequency region is
increasing with increasing SOD. Figure 5.7 (a) depicts the Nyquist-plots for
the impedance response of a consecutively discharged button cell, evaluated
at a SOD of 1%, 24% and 44%, respectively. Here, the same trend for the
arc in the low frequency region is apparent as for the in-house battery.

Figure 5.6 (b) and figure 5.7 (b) show the Bode-plots of the aforementioned
EIS measurements for the in-house battery and the button cell, respectively.
For both battery types, the modulus of the impedance is increasing (slightly
for the button cell) with increasing frequencies above 1 Hz. Moreover, a
minimum value for the phase angle is observed for both batteries around
10^{-1} Hz. The frequency for this minimum is shifting with increasing SOD. As
mentioned for the EIS measurements with various electrolytes, the sluggish
ORR and the oxygen diffusion in the liquid electrolyte are predominant at
this frequency. Thus, it might be deduced that the impedance response of
the air electrode is indirectly affected by a change of the SOD, surprisingly
due to the processes at the zinc electrode.

The trends observed in the low frequency region might be explained
as follows. A volume expansion of the zinc electrode might have caused
a flooding with liquid electrolyte in the GDL of the air electrode, which
subsequently evokes a change in the electrochemical state of the air electrode,

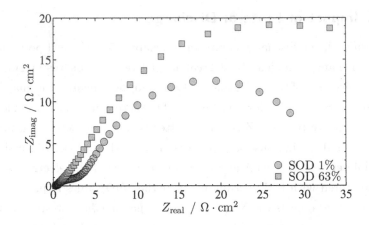

(a)

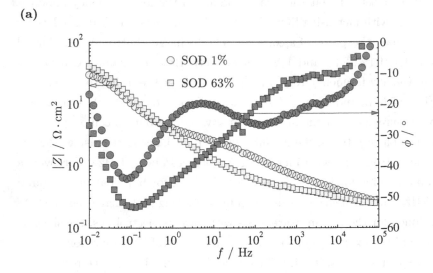

(b)

Figure 5.6.: EIS results at different SOD for the same in-house battery set-up
with 6 M KOH-electrolyte: (a) Nyquist-plot; (b) Bode-plot; data
obtained by [56] and presented in [61].

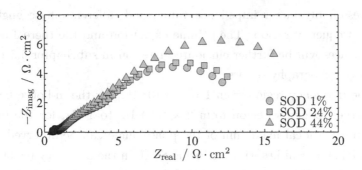

(a)

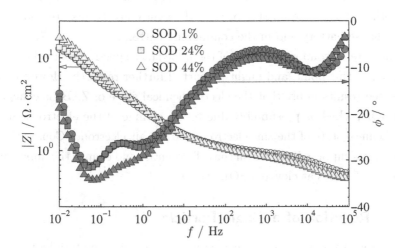

(b)

Figure 5.7.: EIS results at different SOD for the same GP button cell: (a) Nyquist-plot; (b) Bode-plot.

and presumably larger impedance values and different phase angles in the low frequency region. The volume expansion and the thereof arising consequences will be further elucidated later on in subchapter 5.3.1 with the X-ray tomography results.

In the frequency region from 1 Hz to 10^2 Hz, for the in-house battery, and in the frequency region from 2×10^{-1} Hz to 1 Hz, for the button cell battery, a local maximum of the phase angle can be observed. This local maximum vanishes with increasing SOD. In the frequency region from 10^2 Hz to 10^4 Hz, it can be observed for both batteries that the phase angle varies with increasing SOD. This is presumably attributed to the zinc electrode kinetics, since the transport of the reactant at the zinc electrode might be changed. However, this impact is not pronounced in the same way for both batteries, presumably because different constituents are applied in the in-house battery and in the commercial battery.

A definite reason for the shape of the impedance spectra observed can not be given at this point, and might be part of further research. However, the presented results imply that the electrochemical state of ZABs is changing with SOD, which is presumably due to the change of the electrochemical and chemical state of the zinc electrode and the air electrode. Consequently, these effects supposedly diminish the cell potential and change the impedance response of ZABs at elevated SOD values.

5.2.2. Reaction of a Single Particle

To visualize that the expected conversion of the active material zinc takes place in ZABs, a commercially available ZA13-type button cell will be analyzed in the following. It was discharged rheostatically at a condition that is equivalent to a constant current density discharge at $5.4 \text{ mA} \cdot \text{cm}^{-2}$ and monitored with X-ray tomography by researchers of the Helmholtz-Zentrum Berlin [58, 60]. The discharge was stopped at various SOD values so that the X-ray tomography could be conducted.

The conversion of one single zinc particle located near the separator of the button cell investigated, is shown in figure 5.8 with increasing SOD. There, the active material Zn is shown with gray values, oxidized zinc species in the electrolyte are shown in blue color (darkest gray in print version), and greater attenuating additives, such as lead, are shown in red (black in print version). It can be observed that the gray values, indicating Zn, are disappearing in the images while the SOD is increasing, since Zn is presumably electrochemically oxidized. The electrochemical conversion of Zn starts at the outer surface of the particle. The inner regions close to the core of the particle are reacting subsequently. The oxidized zinc species seem to partly dissolve in the electrolyte (appearing as transparent). They might precipitate in the electrolyte and are then transported away from their initial place with increasing SOD, which is apparent as washed out blue (darkest gray in print version) colored areas [60]. At 96.2% SOD, heavy metal additives such as lead (red values (black in print version)), possibly originating from the initial zinc powder (as indicated by [45]), can be observed.

The results obtained for the single particle imply that the electrochemical conversion of Zn to ZnO might be described with a shrinking-core-mechanism [60]. This allows for an even more detailed analysis of the formation of the zinc oxide layer in zinc electrodes during discharge, e.g. with a one-dimensional/one-dimensional (along the zinc electrode and along the particle radius) model-based description, as found for example for lithium-ion batteries.

5.2.3. Volume Expansion

Through-plane center cutouts of the tomography images of the button cell battery for increasing SOD are shown in figure 5.9. In each image, the air electrode is shown at the bottom, the separator is on top of the air electrode, and the zinc electrode is shown on top of the air electrode and the separator. The active material Zn and the battery housing are highly

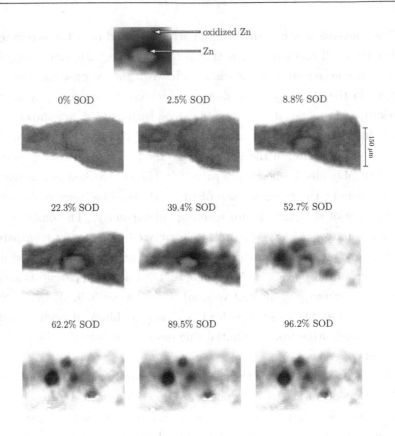

Figure 5.8.: Enlarged three-dimensional reconstruction of a single zinc particle, which is electrochemically converted in the ZA13-type button
cell with increasing SOD. [60] - Published by the PCCP Owner
Societies.

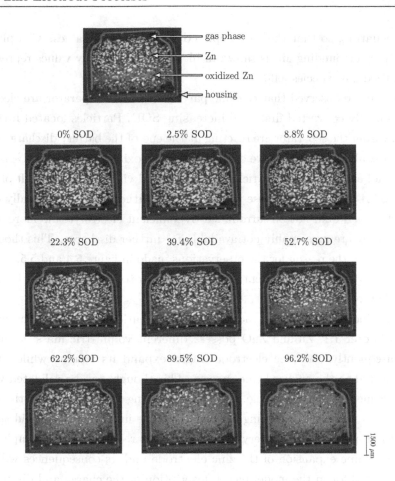

Figure 5.9.: Through-plane center cuts through the ZA13-type button cell
for increasing SOD, achieved by stepwise rheostatic discharge.
The zinc electrode is shown at the top of the image, the air electrode is shown at the bottom, the very thin separator is located
between air electrode and zinc electrode.
[60] - Published by the PCCP Owner Societies.

attenuating, so that they are depicted in bright gray values. Gas phase, i.e. the surrounding air, is shown in black. Mid-level gray values represent oxidized zinc species [60].

It can be observed that the zinc particles near the separator are electro-chemically converted first with increasing SOD. Particles located further away from the separator are reacting at the end of the battery discharge [60]. Almost all zinc particles are converted to zinc oxide species at 96.2% SOD. A small amount of zinc particles is however still visible in the top left of the image at 96.2% SOD. These particles would either not be electrically con-nected, or would not be surrounded by sufficient electrolyte anymore, and thus they are presumably not available for further discharge. This thought might give the reason for the observations made in figure 5.4 and 5.5, where a too high discharge current density prevents to achieve the full battery discharge capacity.

As elucidated with the theoretical consideration for ZAB discharge in subchapter 1.1, Zn and ZnO possess different volumetric mass densities. Consequently, the zinc electrode might expand its volume while Zn is converted to the oxidized zinc species. This thought can be validated with the sequence of images in figure 5.9 for increasing SOD: The top of the zinc electrode is gradually moving upwards, and the initially present void space at the top end of the battery (black values) is reduced almost completely. The volume expansion of the zinc electrode and its consequences will be accounted for in the model-based description of the charge and discharge behavior of ZABs in subchapter 8.2 as such. In addition, the images in figure 5.9 imply that the zinc electrode is compacted. This presumably leads to a local decrease of the zinc electrode porosity [58].

All in all, the results obtained might suggest that if the zinc electrode has no room for expansion in the housing, the liquid electrolyte in the void space between the zinc particles will be transported into the air electrode, and thus into the air holes of the button cell. As a result, the volume expansion of the zinc electrode could change the amount of electrolyte and especially the amount of water in both the zinc electrode and the air electrode significantly. In the images of the button cell investigated, the air electrode is covered with a metal mesh current collector and is close to the metal housing. Both metals are highly attenuating and prevent to monitor the liquid electrolyte, and the pore structure in the air electrode. For this reason, the in-house ZAB set-up for X-ray tomography is used to have a detailed look inside the air electrode in subchapter 5.3.1.

5.3. Air Electrode Processes

In the following, experimental results for processes that can be attributed to the air electrode are presented and discussed. The previous results have already shown that the processes at the zinc electrode indirectly affect the state of the air electrode. Thus, this subchapter directly continues to investigate the impact of the volume expansion at the zinc electrode with increasing SOD on the air electrode. First, the X-ray tomography analysis of the flooding with liquid electrolyte of the air electrode will be presented. To extend the investigation of the electrochemical behavior of the air electrode, the impact of the catalyst at the air electrode on the polarization behavior and the impedance response of ZABs will be shown subsequently.

5.3.1. Flooding of the Air Electrode with Liquid Electrolyte

To analyze the transport of liquid electrolyte in the air electrode, the prepared battery (in-house set-up for X-ray tomography) was discharged and monitored with increasing SOD. The zinc electrode was thereby intentionally prepared with a densely packed zinc electrode and the air electrode was wetted with liquid electrolyte, so that any volume change would evoke a movement of liquid electrolyte from the zinc electrode into the air electrode. The images obtained of the zinc electrode are shown in figure 5.10. The air electrode is not shown in the images but would be located above the top of the images. The active material Zn is highly attenuating, and therefore depicted in bright gray values, while dark gray values represent

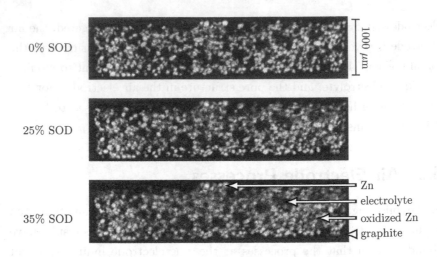

Figure 5.10.: Through-plane cross-sections of the zinc electrode in the in-
house set-up for increasing SOD. The air electrode (not shown)
would be located above the top of the image. The graphite
current collector (large parts not shown) is shown in false color
blue (black in print version).

the electrolyte. Mid-level gray colors represent oxidized zinc species. To
achieve better visual contrast, the background (i.e. the graphite current
collector) is shown in false color blue (black in print version). During
operation, the top of the zinc electrode raises barely, which implies a very
small volume change of the zinc electrode. In addition, the brightest gray
values at 0% SOD are changing gradually to mid-level gray values at 35%
SOD. In detail, a very slight transition of the bright gray values (i.e. pure
zinc) to mid-level gray values (i.e. oxidized zinc) at the edges of the zinc
particles can be observed locally. Although only slightly visible, this implies
that an electrochemical conversion takes place, which is accompanied by a
very small decrease in porosity of the zinc electrode. Thus, less void space
for the liquid electrolyte might be the consequence.

Figure 5.11 shows top views of the GDL of the air electrode of the
very same battery with increasing SOD from 0% to 35%. Thereby, the

tomographic information from ten cut layers along the thickness of the GDL (approximately 250 μm) are summed up to illustrate the color values shown. White and light gray colors thereby represent gaseous species. Dark gray to black colors represent the carbon structure of the GDL. Dark blue (black in print version) colored areas are in false color, and indicate greater X-ray attenuation than for the gas phase and the carbon structure, which can be seen as the liquid electrolyte. It can be observed that the GDL possesses a regular mesh structure of pores in the carbon layer with dark gray and black color with areas containing both gas phase and carbon in light gray values. With increasing SOD, the pores of the GDL are more and more filled with liquid electrolyte, which is visualized by the dark blue (black in print version) areas.

Figure 5.12 shows the results of the X-ray tomography with through-plane cross-sections (side view) of a part of the air electrode GDL (800 μm times 250 μm) of the same battery. The CL is located at the bottom and is the highest attenuating phase in the images, shown in red (black in print version). The surrounding air is located in the top of the images. The colors in these images are false colors, and are chosen to emphasize the change in X-ray attenuation in the GDL pores with increasing SOD. Black values thereby suggest areas with gas phase, and colors from blue (dark gray in print version) to medium red (medium dark in print version), as indicated with the colorbar, indicate areas with more liquid electrolyte than in other areas. As mentioned in subchapter 3.2, the total mass attenuation coefficient obtained for a structure is the weighted sum of the mass attenuation coefficients of the single elements (see equation (3.7)). By implication, rather the change in attenuation in the air electrode should be interpreted than the attribution to a single phase or species. The colors can not be explicitly linked to solely the attenuation of the carbon structure of the GDL or to the liquid electrolyte, but rather to the attenuation of the mixture of both. With increasing SOD, the X-ray attenuation in the GDL, in particular indicated by the medium red (medium dark in print version) values, increases along the thickness of the GDL. This increase can be interpreted as increase of

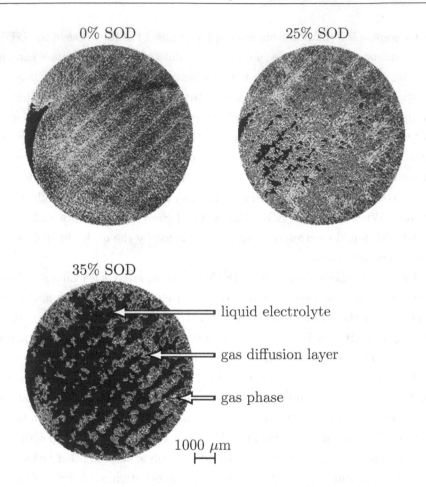

Figure 5.11.: In-plane cross-sections (top view) of the entire air electrode (approximately 250 μm thick) of the in-house set-up at 0% SOD, 25% SOD and 35% SOD. Reprinted from [59] with permission from Elsevier.

the liquid phase in the GDL, which in turn is presumably caused by the earlier reported volume change in the zinc electrode. As observed in figure 5.6, the increasing SOD can impact the EIS response in the low frequency region, which might be caused by an increased liquid electrolyte level as it might represent an increased hindrance for the oxygen diffusion through the pores of the GDL, which are more and more filled with liquid, and as it might represent a change of the electrochemically active area at the CL.

The ZAB for the measurement results shown above could be discharged at a discharge current density of 7.07 mA · cm^{-2} only until 35% SOD because the cell potential was reaching the cut-off voltage. Although there was still sufficient amount of Zn available, as visible in figure 5.10, the discharge could not be continued further than 35% of the SOD. The corresponding discharge curve is shown in figure 5.13. During discharge, the cell potential is initially

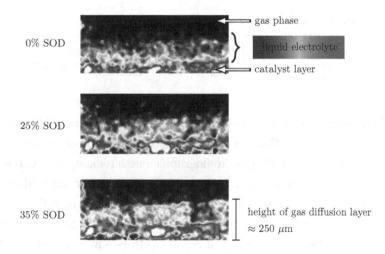

Figure 5.12.: Through-plane cross-sections (side view) of the air electrode for increasing SOD, i.e. enlarged detail of the GDL from figure 5.11. The liquid electrolyte is depicted in false colors, as indicated with the colorbar. The colors are chosen to emphasize the change in X-ray attenuation in the pores with increasing SOD. Reprinted from [59] with permission from Elsevier.

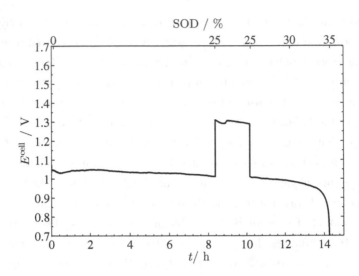

Figure 5.13.: Discharge curve for the prepared in-house battery for the tomography measurements, discharged at a current density of 7.07
mA · cm^{-2}; the SOD at 8.25 hours and 14.25 hours is 25% and
35%, respectively. Reprinted from [59] with permission from
Elsevier.

1.05 V and then slightly decreases linearly with operation time. The peak in
cell potential at about 8.25 hours of operation is due to zero current condition,
which was maintained during the tomographic measurement, and represents
the battery's OCP of about 1.30 V at 25% SOD. After approximately
14.25 hours of operation, which corresponds to 35% SOD, the cell potential
decays drastically and the constant current density discharge is terminated
at 0.70 V. The observed transport of liquid electrolyte into the pores of the
air electrode GDL has serious consequences for the cell potential obtained:
due to an increase in the degree of flooding in the GDL pores, the diffusion
of oxygen to the reaction zone at the CL is hindered; some transport paths
in the GDL might even be blocked completely. Since the cell current, and
as such the oxygen consumption in the air electrode, are set constant, the
oxygen concentration in the CL decreases with advancing operation time.

This presumably leads to oxygen starvation in the CL, and finally to a sudden decay in cell potential and consequently to an early end-of-life for the battery [59].

All in all, the presented combination of discharge curve measurement and X-ray tomography enables to reveal an in-depth insight of the processes occurring at the air electrode.

5.3.2. Catalyst Impact

The catalyst applied at the air electrode is another important parameter for the performance of ZABs besides the previously investigated SOD and liquid electrolyte redistribution. The catalyst directly influences the rate of reaction and thus the electrochemical performance of the air electrode. To reveal the impact of the catalyst applied at the air electrode, the following results are presented and discussed.

The in-house set-up was operated with a commercially available GDL with Pt/C-based CL and an in-house GDL with perovskite catalyst as air electrode, respectively. Polarization curves were measured to evaluate the impact of the respective catalyst. The results obtained are shown in figure 5.14. To supplement the findings, EIS results for the same batteries are shown afterwards.

It can be observed that the OCP for the Pt-based catalyst is 0.24 V larger than for the perovskite catalyst. In addition, the initial drop of cell potential in the activation region of the polarization curve is slightly lower for the battery with Pt-based catalyst (0.10 V from 0 mA $\cdot$ cm^{-2} to 1.90 mA $\cdot$ cm^{-2}) compared to the battery with perovskite catalyst (0.16 V for the same current drop). Both aforementioned observations can be attributed to the larger activity of Pt for oxygen reduction (see [69] and [70]) compared to the perovskite catalyst. The cell potential obtained for the curves with perovskite catalyst is in line with experimental data from ZABs with the same catalyst at the air electrode [71].

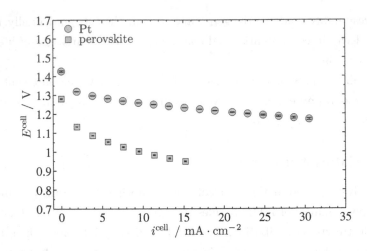

Figure 5.14.: Polarization curve measurements for in-house batteries with Pt-catalyst and perovskite-catalyst at the air electrode with 6 M KOH-electrolyte; measurements carried out by [56].

Furthermore, the slope of the cell potential with increasing cell current density in the quasi ohmic region is lower for the curve with Pt. This might be attributed to a greater ohmic resistance for the in-house GDL with the perovskite catalyst. The general offset in cell potential in the quasi ohmic region might also be explained by the overpotential of the air electrode, (usually due to the different reaction kinetics with different catalyst), or the initial offset in OCP observed.

The impedance response of an in-house battery with 6 M KOH-electrolyte and with perovskite catalyst and Pt catalyst is shown in figure 5.15, respectively. It can be observed that the impedance values in the low frequency region are much larger for the pervoskite-type catalyst than for the Pt catalyst. This leads to the presumption that the low frequency region is affected by the air electrode. Since, the air electrode with Pt catalyst possesses a larger activity for ORR than the air electrode with perovskite-type catalyst (see [69] and [70]), and additionally a different composition and structure, a

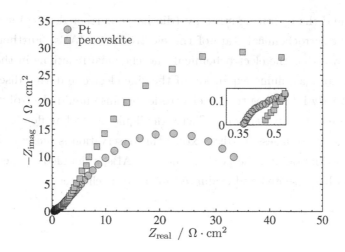

Figure 5.15.: EIS results for batteries with Pt-catalyst and perovskite-catalyst with 6 M KOH-electrolyte, $C_{\mathrm{Ah,initial}} = 343\ \mathrm{mA \cdot h}$ and $342\ \mathrm{mA \cdot h}$, respectively; measurements carried out by [56].

much larger battery impedance in the low frequency region is expected for the air electrode with perovskite-type catalyst applied in this thesis.

The research on catalysts aims to achieve larger operating cell potential during charge and discharge (see [43] and [21]). Since Pt/C-based electrodes are not stable enough during OER, new catalysts at the air electrode also need to be long-term stable and highly active for both the ORR and OER. The results shown in this subchapter therefore emphasize the need for improved catalysts for the air electrode of ZABs.

In summary, the experimental analysis presented in this chapter gives a better understanding of the reactions at zinc and air electrode, and of the transport of electrolyte within ZABs. Moreover, the analysis has shown that the processes at both electrodes are interconnected and that they largely influence each other. In particular, the experimental results have revealed that the battery state (electrolyte composition, SOD, catalyst applied) and the operating strategy (discharge current density) strongly affect the electrochemical state of the electrodes in ZABs. Furthermore, it was shown that the electrochemical and chemical reactions in the zinc electrode cause a volume expansion of the zinc electrode during discharge. This indirectly influences the air electrode, because liquid electrolyte can be pushed into the air electrode. There, the liquid electrolyte floods the gas diffusion layer. To assess to what extent this observation is relevant for the oxygen distribution and the performance of ZABs, a detailed air electrode model will be presented and evaluated in the next chapter.

6. Detailed One-Dimensional Air Electrode Model

As observed in the previous chapter, the volume expansion of the zinc electrode during ZAB discharge can flood the pores of the GDL with liquid electrolyte. A model-based analysis is applied in this chapter of the thesis to assess the implications of flooding on the air electrode overpotential, and the oxygen distribution within the air electrode. This chapter is based on the publications [72], [73], and [74].

6.1. Model Description

In the following, a one-dimensional model of an air electrode, containing in particular partial differential equations for species concentrations, is presented. The model description is given in general terms, so that it can be adapted or extended for air electrodes in other electrochemical systems.

The air electrode considered in this chapter is schematically shown in figure 6.1. This air electrode is situated in the one-dimensional domain $\Omega(x)$, is composed of a porous GDL with the thickness δ_{GDL}, and faces on one side a surrounding gas phase and on the other side an attached CL. The GDL is considered to be partly filled with liquid electrolyte (i.e. KOH-solution for ZABs) and with air or pure oxygen.

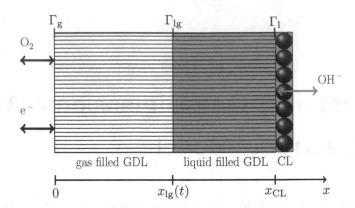

Figure 6.1.: Schematic of the air electrode considered, with gas diffusion
layer (GDL), catalyst layer (CL), and the parts filled with liquid
electrolyte (light gray) and gas phase (white). The location x_{lg} of
the interface between gas and liquid, Γ_{lg}, can change with time.
The position of the catalyst layer is at x_{CL}. It marks the end of
the considered domain $\Omega(x)$ for the analysis of the air electrode.
Reprinted from [74] with permission from Elsevier.

Molar Balances and Charge Balance

Concentrations for the species $j = O_2$ in the gas phase, and $j = O_2, OH^-$
in the liquid phase of the air electrode, are calculated with the following
molar balance in concentration form with a source or sink term, as proposed
by [75]:

$$\frac{\partial c_j(t,x)}{\partial t} = \nabla\left(-D_{j,h}^{\text{eff}} \cdot \nabla c_j(t,x)\right) + Q_j(t,x) \tag{6.1}$$

where, c_j is the molar concentration of species j, $D_{j,h}^{\text{eff}}$ is the effective diffusion coefficient for the diffusion of a species through the fluid h, which
is corrected for the porosity of the GDL, ε, as presented in subchapter 1.1
in equation (1.5). The diffusion coefficients $D_{O_2,\text{KOH}}^{\text{eff}}$ and $D_{OH^-,\text{KOH}}^{\text{eff}}$ are
implemented as a function of c_{OH^-} in this chapter with data by [76] and [77],
respectively. The respective expression are given in the appendix in equations (A.1) and (A.2).

$Q_j(t, x)$ is a source or sink term for the species j, which is given by the chemical or electrochemical reactions in which the species participates. For this chapter, solely the electrochemical reaction (IV), the ORR/OER, is considered to take place. It is assumed to occur directly at the left-hand-side end of the catalyst layer at the position x_{CL}. Consequently, Q_j becomes zero for all $x < x_{CL}$. Directly at x_{CL}, Q_j is applied as the product of the volume specific reaction rate, $r(t)/A\cdot\delta_{CL}$, and the stoichiometric coefficient of species j in reaction (IV), ν_j, so that:

$$Q_j(t, x) = \begin{cases} 0 & \text{if } x < x_{CL} \\ \nu_j \cdot \dfrac{r(t)}{A \cdot \delta_{CL}} & \text{if } x = x_{CL} \end{cases} \tag{6.2}$$

where A is the cross-sectional area of the electrode, and δ_{CL} is the thickness of the catalyst layer. Thereby, the reaction rate in the CL is uniform and averaged in the volume (cross-sectional area times thickness) of the CL.

$r(t)$ is expressed in this chapter by a Butler-Volmer expression, and represents the reaction rate of reaction (IV), the ORR/OER at the air electrode, as proposed by [71]:

$$\begin{aligned} r(t) = \quad & k_{\text{forw}} \cdot \frac{c_{H_2O}}{c^{\text{ref}}} \cdot \left(\frac{c_{O_2}^*}{c^{\text{ref}}}\right)^{\frac{1}{2}} \cdot \exp\left(\frac{-(1-\alpha)\cdot \mathbf{F}}{\mathbf{R}\cdot T} \cdot \eta^{\text{air}}\right) \\ & -k_{\text{backw}} \cdot \left(\frac{c_{OH}}{c^{\text{ref}}}\right)^2 \cdot \exp\left(\frac{\alpha \cdot \mathbf{F}}{\mathbf{R}\cdot T} \cdot \eta^{\text{air}}\right) \end{aligned} \tag{6.3}$$

where η^{air} is the overpotential of the air electrode, $c_{O_2}^*$ is the concentration of dissolved oxygen directly at the CL at x_{CL}, c^{ref} is a reference concentration of $1 \text{ mol} \cdot \text{dm}^{-3}$, k_{forw} and k_{backw} are the reaction rate constants for the forward and backward reaction of reaction (IV), respectively. $\mathbf{F}$ is Faraday's constant, $\mathbf{R}$ is the universal gas constant, α is the transfer coefficient for the electrode reaction, and T is the temperature considered for the air electrode.

To account for the activation losses due to polarization of the air electrode, a charge balance is applied to describe the time-dependent behavior of the overpotential η^{air} at the electrode, so that:

$$\frac{\mathrm{d}\eta^{\text{air}}}{\mathrm{d}t} = -\frac{1}{C_{\text{CL}}} \cdot \left(i_{\text{cell}} - 2 \cdot \mathbf{F} \cdot \frac{r(t)}{A} \right) \tag{6.4}$$

where C_{CL} is the double layer capacitance of the electrode, and i_{cell} is the cell current density. i_{cell} is a time-dependent input value for the air electrode, which enables to analyze pulse-current operation for the air electrode later on in this chapter.

Volume Change

The liquid volume in the air electrode can change due to a multitude of impacts during operation. One example is shown by the experimental results in subchapter 5.3; the volume of the liquid electrolyte in the GDL can increase due to a volume expansion of the zinc electrode. Whatever the cause might be for the liquid volume to change, as a consequence the position x_{lg} of the interface between liquid and gas phase, Γ_{lg}, in the GDL will change likewise.

To account for a variable volume of liquid electrolyte in the air electrode, an ordinary differential equation for the change of the volume V_{l} in the liquid filled part of the air electrode is applied, so that:

$$\frac{\mathrm{d}V_{\text{l}}(t)}{\mathrm{d}t} = A \cdot \frac{\mathrm{d}x_{\text{lg}}(t)}{\mathrm{d}t} \tag{6.5}$$

Thereby the cross-sectional area A of the liquid filled part of the GDL is considered to be constant. Thus, the change in liquid volume $\mathrm{d}V_{\text{l}}(t)/\mathrm{d}t$ determines the position of the interface between liquid and gas phase Γ_{lg}. As a consequence, $x_{\text{lg}}(t)$ can deviate from its initial value $x_{\text{lg}}(t = 0)$ during operation.

Model Domains, Initial Conditions, and Boundary Conditions

The governing equations are considered for the entire space-domain $\Omega(x)$ of the air electrode. The domain contains a gas filled and a liquid filled sub-domain, which are Ω_g and Ω_l, respectively. Each sub-domain is confined by two interfaces that are indicated with Γ. Thus, there are four interface boundaries in the air electrode (see also figure 6.1): One outer boundary interface to the ambient air, Γ_g, one inner boundary interface, Γ_l, and two inner boundary interfaces, Γ_{gl} and Γ_{lg}, at the same position in between Ω_l and Ω_g.

Four initial conditions for the species O_2 and OH^- in the partial differential equation (6.1) are chosen, so that:

$$c_{O_2}(0,x) = \hat{c}_{O_2} \qquad\qquad \forall\, x \in \Omega_g \qquad\qquad (6.6)$$

$$c_{O_2}(0,x) = \hat{c}_{O_2}^* \qquad\qquad \forall\, x \in \Omega_l \qquad\qquad (6.7)$$

$$c_{OH}(0,x) = 0 \qquad\qquad \forall\, x \in \Omega_g \qquad\qquad (6.8)$$

$$c_{OH}(0,x) = \hat{c}_{OH} \qquad\qquad \forall\, x \in \Omega_l \qquad\qquad (6.9)$$

where * denotes dissolved oxygen, and ˆ indicates a value at $t = 0$. The concentration of dissolved oxygen is given by the solubility according to Henry's law, so that:

$$c_{O_2}^* = p_{O_2} \cdot H \qquad\qquad (6.10)$$

where H is Henry's law coefficient, which is a function of c_{OH}. It is calculated as given in equation (A.12) in the appendix. The implications of applying Henry's law, a thermodynamic equilibrium expression, in a time-dependent equation are discussed in more detail in the appendix in subchapter A.1.

The initial conditions for the ordinary differential equations (6.4) and (6.5) determine the initial overpotential at the air electrode, and the initial filling level of liquid electrolyte in the GDL. In general, they are:

$$\eta^{air}(0) = \hat{\eta}^{air} \qquad\qquad (6.11)$$

$$V_1(0) = \hat{V}_1 = A \cdot \delta_{\mathrm{lg}}(t = 0) \tag{6.12}$$

where $\delta_{\mathrm{lg}}(t = 0)$ is the initial thickness of liquid film present in the GDL, ranging from x_{CL} to x_{lg}.

Two Dirichlet and two von-Neumann boundary conditions are chosen for the sub-domains to represent the conditions of the adjacent surrounding air and the battery separator, so that:

$$c_{O_2}(t, x) = \hat{c}_{O_2} \qquad \forall\, x \in \Gamma_g\, \forall\, t \tag{6.13}$$

$$\frac{\partial c_{O_2}}{\partial x} = 0 \qquad \forall\, x \in \Gamma_l\, \forall\, t \tag{6.14}$$

$$\frac{\partial c_{OH}}{\partial x} = 0 \qquad \forall\, x \in \Gamma_g\, \forall\, t \tag{6.15}$$

$$c_{OH}(t, x) = \hat{c}_{OH} \qquad \forall\, x \in \Gamma_l\, \forall\, t \tag{6.16}$$

Equations (6.1), (6.4), and (6.5) are the resulting set of partial and ordinary differential equations. They are first discretized in space with a moving grid finite volume method, as shown in more detail in the appendix in subchapter A.1 (equation (A.4)). Then, they are solved in time with the Matlab ode15s solver, a variable order numerical differentiation algorithm for stiff systems [78].

6.2. Implementation of Flooding and Pulse-Current Operation

After the description of the basic model equations, two modifications are introduced to qualitatively analyze the oxygen distribution and the over-potential of the air electrode. The model parameters applied for the two modifications are given in table A.1. The two modifications are as follows:

- First, a constant increase of liquid electrolyte volume of $\mathrm{d}V_l(t)/\mathrm{d}t = 1 \times 10^{-12}\ \mathrm{m}^3 \cdot \mathrm{s}^{-1}$ is implemented for the differential equation (6.5). This shall represent a flooding mechanism, for which the position of the interface of gas and liquid $x_{\mathrm{lg}}(t)$ is continuously shifted towards the exterior of the electrode. This modification is implemented to represent the experimentally observed increase of liquid electrolyte in the GDL during galvanostatic operation (see subchapter 5.3.1).

- Second, a pulse-current operation is chosen as model input. Pulse-current operation implies a fluctuating cell current as input value for the air electrode. It is specified in this thesis by the parameters pulse-current density, i_{pulse}, in $\mathrm{mA} \cdot \mathrm{cm}^{-2}$, pulse duration in seconds, and recovery time in seconds, which is a resting period for the air electrode at a current density of $0 \ \mathrm{mA} \cdot \mathrm{cm}^{-2}$.

The pulse-current operation is chosen as air electrode input since it can prevent dendrite formation (item 4 on the list of drawbacks for ZAB operation in subchapter 1.2) for ZABs [79]. However, the pulse-current response of the air electrode in ZABs, has not been investigated in detail so far, and resembles a promising alternative to the conventionally applied constant current density discharge.

6.3. Simulated Overpotential and Oxygen Distribution

Figure 6.2 shows the transient response of the air electrode for the first modification to account for a constant increase of liquid volume in the GDL, i.e. flooding. Thereby, figure 6.2 (a) depicts the overpotential of the air electrode during a constant current density discharge with $2 \ \mathrm{mA} \cdot \mathrm{cm}^{-2}$ while the liquid volume in the GDL increases steadily with $\mathrm{d}V_{\mathrm{l}}(t)/\mathrm{d}t = 1 \times 10^{-12} \ \mathrm{m}^3 \cdot \mathrm{s}^{-1}$. In addition, the overpotential response to the same current density without increasing liquid volume ($\mathrm{d}V_{\mathrm{l}}(t)/\mathrm{d}t = 0 \ \mathrm{m}^3 \cdot \mathrm{s}^{-1}$), i.e. no flooding, is shown as a baseline in the same figure. It can be observed that the overpotential for constant flooding first decreases linearly from -0.34 V to -0.36 V during 5 hours of operation, and then decreases more rapidly within 2 hours to -0.44 V. This behavior can be explained as follows. Since the flooding and the thickness of the liquid phase in the GDL increase constantly, the diffusion pathway through the liquid filled GDL to the CL increases for the dissolved oxygen. Consequently, the concentration of dissolved oxygen at the CL would decrease. Thus, the reaction rate would be diminished, whereas the electrode current density is kept constant. As a consequence, the air electrode overpotential has to decrease according to equation (6.3) and (6.4).

The concentration of dissolved oxygen at the catalyst layer, directly at x_{CL}, is depicted in figure 6.2 (b). In the case of flooding, the oxygen concentration decreases linearly with increasing operation time from 7.30×10^{-3} to 5.25×10^{-3} $\mathrm{mol \cdot l^{-1}}$. For the simulation with no flooding, the oxygen concentration remains constant during the entire operation. The linear decline of the oxygen concentration at the CL with operation time during flooding helps to explain the progression of the overpotential in figure 6.2 (a). η^{air} and $c_{O_2}^*$ at x_{CL} are linked via an exponential expression for the reaction rate and the charge balance for the air electrode (see equation (6.3) and (6.4), respectively). Thus, a linear decline of the oxygen concentration at the catalyst layer evokes an exponential decrease of the overpotential of the air electrode with constant flooding.

From the results obtained, the following implication can be derived for the operation of a full cell ZAB. In general, the cell potential of ZABs is primarily determined by the overpotential at the air electrode [71], so that a decrease of the oxygen concentration at the CL can severely decrease the air electrode overpotential and thus can diminish the overall cell potential of ZABs. For instance, the decreasing cell potential observed in the discharge curve for the X-ray tomography set-up in figure 5.13, might be attributed to an increased overpotential due to a flooded air electrode.

In sum, the X-ray tomography results and the model-based results indicate that the latter might be the reason for the early-end-of-life observed for the particular battery. However, it can not be fully excluded that the overall cell potential was also affected by other battery components such as zinc

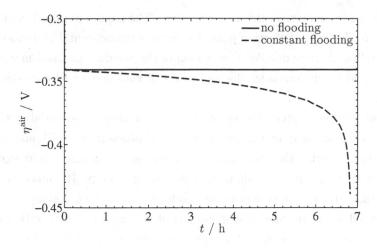

(a)

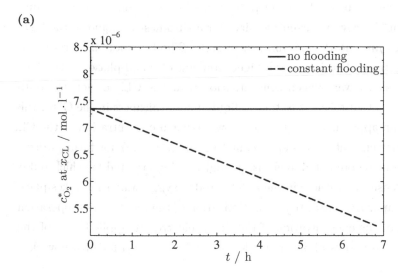

(b)

Figure 6.2.: Simulation results for the transient response of the air electrode:
(a) overpotential; (b) oxygen concentration at the catalyst layer.
Constant current density discharge with $2\,\mathrm{mA \cdot cm^{-2}}$, for the
case that the liquid volume in the GDL is increasing with a
constant rate of $\mathrm{d}V_{\mathrm{l}}(t)/\mathrm{d}t = 1 \times 10^{-12}\,\mathrm{m^3 \cdot s^{-1}}$, and for no flooding.

electrode and separator in this measurement. The results obtained in this chapter can be seen as starting point for further experimental and model-based studies. A more detailed description of the flooding mechanism with updated model equations is feasible and might be considered for future work.

For the previously presented simulation results, the overpotential at the air electrode is strongly influenced by the liquid phase in the GDL and the oxygen concentration therein. Another parameter to influence the oxygen distribution in the air electrode is the set current density. Its impact will be investigated with the pulse-current modification in the following.

Figure 6.3 shows the spatial distribution of the oxygen concentration in the liquid filled domain $\Omega_l(x)$ for a pulse-current input (1.71 mA $\cdot$ cm^{-2} pulse-current density, 1 second pulse duration, 6 seconds recovery time, 0 mA $\cdot$ cm^{-2} recovery current density) for a thickness of liquid in the GDL of (a) 0.25 μm and (b) 1.4 μm, respectively. The two plots illustrate that an air electrode that possesses a higher amount of liquid phase in the GDL, will possess a lower oxygen concentration near the CL. In fact, the data points reveal that a 5.6 times thicker liquid phase thickness in the GDL, will result in an approximately 400 times lower oxygen concentration at the CL. These pulse-current results are meant to give an outlook for future work.

Based on the presented analysis, it might be investigated at which pulse-current density and where in the air electrode oxygen starvation takes place. With the approach presented, an optimization of the pulse-current operation strategy, of the concentration of the liquid electrolyte applied, and of the structural properties of air electrodes might be achieved for future work.

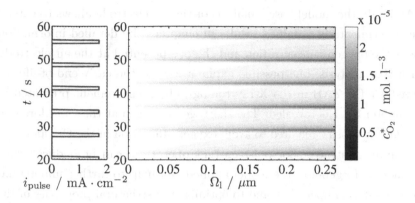

(a)

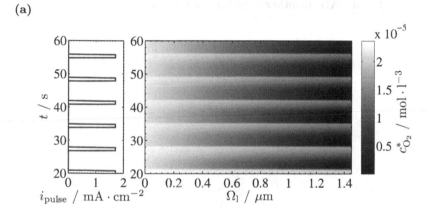

(b)

Figure 6.3.: Spatial oxygen distribution within the liquid phase sub-domain Ω_l of the air electrode for pulse-current operation with 1.71 $mA \cdot cm^{-2}$, for a pulse duration of 1 second, recovery time of 6 seconds at a current density of 0 $mA \cdot cm^{-2}$: (a) simulation result with x_{lg} at 0.25 μm (low flooding); (b) simulation result with x_{lg} at 1.4 μm (high flooding). Reprinted from [74] with permission from Elsevier.

All in all, the model-based analysis of the air electrode allows to extend the experimentally obtained results in chapter 5 by detailed information about the oxygen concentration and the overpotential at the air electrode. With this analysis, one possible explanation for the early-end-of-life observed for the ZAB in the X-ray tomography measurement presented in subchapter 5.3.1 is revealed: The shortage of oxygen at the CL. Moreover, the current density required to avoid oxygen starvation in the CL, and even an operating strategy for pulse-current operation, can be deduced with the model. In general, the model applied can help to further understand processes during operation, and to optimize the structural properties of the air electrodes in ZABs or other metal-air batteries.

Part 2 – Identifying Factors for Long-Term Stable Operation

7. Theoretical Considerations on Air-Composition Impact

As mentioned in subchapter 1.1, the main advantage of ZABs, the superior theoretical energy density, originates from the fact that O_2 is not stored within the cell but is taken from the surrounding air. This requires on the one hand an open air electrode concept for ZABs, but makes on the other hand the entire system susceptible to the surrounding air [1]. Passaniti et al. present a comprehensive overview, and state experimental results on the impact of the surrounding air on ZAB button cells [29]. However, the overview is not given for the operation of electrically rechargeable ZABs.

Especially for long operating times of electrically rechargeable ZABs, i.e. for numerous charge and discharge cycles, the composition of the surrounding air with its relative humidity, carbon dioxide content, oxygen content and temperature can presumably change the state inside the ZAB cell significantly. To qualitatively assess their relevance for ZAB operation, theoretical considerations about the impact of these parameters will be presented with the help of literature sources in this chapter. The relations and equations explained and presented in the following are the basis for the quantitative model-based analysis of electrically rechargeable ZABs, which will be presented in chapter 8.

7.1. Relative Humidity

One impact on the half-open ZAB might be the relative humidity in the surrounding air. Alkaline liquid electrolytes tend to take up or release

gaseous water from or to the surrounding [80]. The extent of the loss
or gain of water depends on the state of the electrolyte, which is defined
by its thermodynamic properties such as chemical potential, respectively
concentration, and temperature. The loss or gain of the solvent water in the
electrolyte at the air electrode may cause a significant depletion or increase
of the total water amount in the battery, and thus might lead in the end to
complete battery failure.

Balej measured the vapor pressure that will adjust above a KOH-solution
as a function of temperature and KOH-molarity of the electrolyte [80].
With the data given, it is possible to calculate the relative humidity in
the surrounding air, which is required to keep the water vapor directly
above the electrolyte in equilibrium with the surrounding water vapor. To
adjust equilibrium between both, the molar fraction directly above the
electrolyte phase, $y_{H_2O(g)}^{sat}$, and the molar fraction of the gaseous water in
the surrounding environment, $y_{H_2O(g)}^{env}$, have to be equal, so that:

$$y_{H_2O(g)}^{sat} = y_{H_2O(g)}^{env} \tag{7.1}$$

$y_{H_2O(g)}^{sat}$ is thereby given by the data from Balej [80], and $y_{H_2O(g)}^{env}$ can be
expressed with the relative humidity in the surrounding, RH^{env}, as follows:

$$y_{H_2O(g)}^{env} = RH^{env} \cdot \frac{p_{H_2O(g)}(T^{env})}{p_{atm}} \tag{7.2}$$

where p_{atm} is the atmospheric pressure, and $p_{H_2O(g)}$ is the saturation partial pressure of gaseous water, which depends on the temperature in the
surrounding, T^{env}. It is calculated with temperature dependent correlations
given by Stull [81] in this thesis, as given in equation (A.47) in the appendix.

Combining equation (7.1) and equation (7.2), rearranging, and using the
relations given by [80], yields the relative humidity required to adjust the
equilibrium, RH^{equi}:

$$y_{H_2O(g)}^{sat} = y_{H_2O(g)}^{env} = RH^{equi} \cdot \frac{p_{H_2O(g)}}{p_{atm}} \tag{7.3}$$

$$\log y_{H_2O(g)}^{sat} =$$

$$(b_1 \cdot \tilde{c}_{KOH} + b_2 \cdot \tilde{c}_{KOH}^2 + b_3 \cdot \tilde{c}_{KOH}^3$$

$$+ (1 + b_4 \cdot \tilde{c}_{KOH} + b_5 \cdot \tilde{c}_{KOH}^2 + b_6 \cdot \tilde{c}_{KOH}^3)$$

$$\cdot (b_7 + \frac{b_8}{T} + b_9 \cdot \log T + b_{10} \cdot T)) / (b_7 + \frac{b_8}{T} + b_9 \cdot \log T + b_{10} \cdot T)$$

$$(7.4)$$

$$RH^{equi} = \frac{y_{H_2O(g)}^{sat} \cdot p_{atm}}{p_{H_2O(g)}} \qquad (7.5)$$

with b being empirical parameters given by [80], and stated in the appendix in equations (A.37) to (A.46). Equation (7.4) is valid for KOH-solutions from 0 to 18 M and from 273 to 573 K [80]. Therein, $\tilde{c}_{KOH}$ is the liquid electrolyte molality. In this thesis, the applied molal concentrations are converted to molar concentrations as given in the appendix in equation (A.48).

Equation (7.5) is used in the following to assess whether the electrolyte will take up or lose water vapor as a function of relative humidity and temperature of the surrounding environment. Figure 7.1 (a) indicates which RH^{env} and which T need to be established to reach equilibrium between the water vapor above the electrolyte and the surrounding water vapor at a given KOH-molarity. Every set of RH^{env} and T established that is not on the curve plotted for a given KOH-molarity, implies water loss (left of the curve) or water gain (right of the curve) for the electrolyte solution, which can either lead to drying out or flooding of the entire battery. Both cases, drying out and flooding, are not favorable for ZAB operation because they can lead to an early end-of-life, as the CL is either not in contact with electrolyte, or as the CL is covered by an excess of electrolyte, as shown with the X-ray tomography measurements in figure 5.11 and 5.12.

The gray rectangle in figure 7.1 (a) highlights the region where room temperature can be found, i.e. at $T^{env} = 298$ K. This temperature is used to

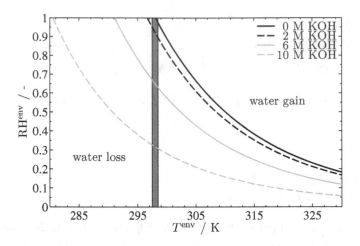

(a)

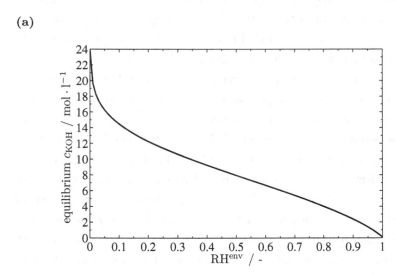

(b)

Figure 7.1.: (a) Equilibrium lines for the exchange of gaseous water between electrolyte and surrounding(see equation (7.5)), T of electrolyte 298 K; (b) KOH-molarity, c_{KOH}, that will adjust in equilibrium for the electrolyte when exposed to RH^{env} at $T^{env} = 298$ K.

depict the graph in figure 7.1 (b). It shows the equilibrium KOH-molarity as a function of RH^{env} at $T^{env} = 298$ K. The shown relation can be used to predict the KOH-molarity that will adjust in chemical equilibrium when the KOH-solution is exposed to a certain RH^{env}. This plot will help to understand the simulation results obtained later on. Already at this point, an important aspect for ZAB operation can be seen from the graph. For RH-values close to zero, the KOH-molarity that will adjust increases drastically and the maximal solubility for KOH of 15 $mol \cdot l^{-1}$ (see figure 7.5) could be reached, which is not favorable for ZAB operation since KOH-salts might precipitate in the bulk of the electrolyte or locally in the GDL.

It is to be noted that for these considerations the temperature T of the electrolyte is set to 298 K. Cooling or heating the electrolyte (or eventually the ZAB) will change the state of the water vapor above the electrolyte and thus will change the outcome of figure 7.1 (b).

In addition, any change in KOH-molarity, for example due to carbonation or due to dilution caused by water gain, will change the equilibrium and could possibly lead to a subsequent loss or a gain of water as well. For a set surrounding temperature of 298 K and a RH of 0.65, only a KOH-solution around 6 $mol \cdot l^{-1}$ would be in equilibrium with the surrounding. Since usually the electrolyte in ZABs is not adjustable during operation, the electrically rechargeable ZABs is strongly affected by changes in RH^{env} and T^{env} during operation, so that on dry days the ZAB might dry out and that on humid and hot days, the ZAB might be flooded with water. This issue will be further elucidated for the charge and discharge cycling of ZABs by the simulation results in subchapter 9.2.

7.2. Carbon Dioxide

Another impact factor might be the carbon dioxide concentration in the surrounding air. Liquid electrolytes, such as KOH-solution, react with carbon dioxide and form carbonate species. Due to the elevated pH of the KOH-electrolyte, the predominant carbonate species is the carbonate ion

(CO_3^{2-}) [82], so that the following irreversible reaction will be taken into account for this thesis:

$$CO_2(\text{diss}) + 2\,OH^- \xrightarrow{r_V} CO_3^{2-} + H_2O \qquad\qquad (V)$$

Reaction (V), also known as the carbonation reaction, affects the long-term stable operation of ZABs, which was shown experimentally: The negative effect was explained by Drillet et al. by salt precipitation after electrolyte carbonation in the air electrode, which subsequently blocked gas diffusion pores and cuts off oxygen supply at the air electrode [83]. In addition, the reaction kinetics of the zinc electrode might be affected by carbonates, which is indicated by single zinc electrode measurements for a mixture of KOH-K_2CO_3-electrolyte [84]. Moreover, the ionic conductivity of the electrolyte is diminished due to a replacement of OH^- by CO_3^{2-} if carbonation takes place [85]. Only the latter is included in this thesis.

The rate of the carbonation reaction strongly depends on the OH^--molarity of the electrolyte [82]. Adding CO_3^{2-} to the electrolyte might slow down the carbonation reaction while still sufficient ionic conductivity is maintained [57]. In general, the reaction rate constant of the carbonation reaction depends on the reaction temperature, on the nature, and the ionic strength of the electrolyte applied (see [86] or [82] for a comprehensive overview).

Figure 7.2 depicts the reaction rate constant k_V as a function of OH^--concentration. It can be observed that the greater the concentration of OH^- is, the greater the reaction rate constant is. The other way around, the carbonation reaction will be slowed down when only small amounts of OH^- are present. The reaction rate r_V of the carbonation reaction again strongly depends on the concentration of OH^-, c_{OH}^{air}, in the liquid electrolyte of the air electrode, as introduced by Pohorecki and Moniuk [87]:

$$r_V = k_V \cdot V_{\text{electrolyte}}^{air} \cdot c_{OH}^{air} \cdot c_{CO_2}^{*} \qquad\qquad (7.6)$$

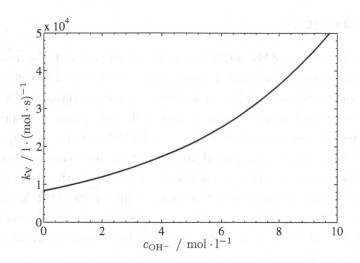

Figure 7.2.: Reaction rate constant for the carbonation reaction (V) as a
function of OH⁻-concentration; data given by Tseng and Ho [86]
and calculated with equation (A.36) at $T = 298$ K.

where $c^*_{CO_2}$ represents the molar concentration of dissolved carbon dioxide in
the liquid electrolyte bulk of the air electrode, and $V^{air}_{electrolyte}$ is the volume
of the liquid electrolyte in the air electrode.

Moreover, reaction (V) implies that one water molecule is produced due
to the deleterious carbonation reaction. This could have a beneficial side
effect if loss of water, as described in subchapter 7.1, takes place during
ZAB operation; the depletion of water could be diminished on the one hand
by sacrificing ionic conductive OH⁻ on the other hand.

All in all, the aforementioned deleterious effects of the carbonation reaction can significantly impact the operation stability of ZABs, which are
operated with the standard KOH-electrolyte molarity around 6 mol·l⁻¹.
This issue will be investigated in more detail with the model-based analysis
in subchapter 9.

7.3. Oxygen

Another influence on ZABs might be the partial pressure of the surrounding or supplied oxygen, which has a strong impact on the solubility of oxygen in the liquid electrolyte. Figure 7.3 shows the concentration of dissolved oxygen for the partial pressure of oxygen in air and for one atmosphere of oxygen partial pressure, and as a function of KOH-molarity. It is apparent that the larger the oxygen partial pressure above the electrolyte solution is, the larger the concentration of dissolved oxygen in the electrolyte is; it is approximately one order of magnitude higher for 0.101 MPa than for 0.021 MPa. In addition, an increase in KOH-molarity decreases the concentration of dissolved oxygen in the electrolyte exponentially; the larger the KOH-molarity, the lower the concentration of dissolved oxygen. This is due to the presence of larger amounts of K^+ and OH^-, which represent a barrier for the oxygen molecule to dissolve into the electrolyte solution. An enhanced oxygen solubility in the electrolyte is desirable for an improved discharge performance of ZABs since a greater oxygen concentration will be beneficial for the ORR at the air electrode.

Limiting Current Density due to Oxygen Transport Limitation

A concentration of dissolved oxygen approaching zero at the CL can be correlated to a maximal current density that can be drawn from the ZAB during discharge. This maximum of current density is strongly determined by the properties of the gas and the electrolyte applied, and will be outlined in the following.

Since, the diffusion coefficient for dissolved oxygen in KOH, $D_{O_2,KOH}$, is in general much lower than the diffusion coefficient for gaseous oxygen in air or N_2 (see table A.3), the diffusion of gaseous oxygen might not be considered as limiting. Standard electrochemistry textbooks, e.g. [88],

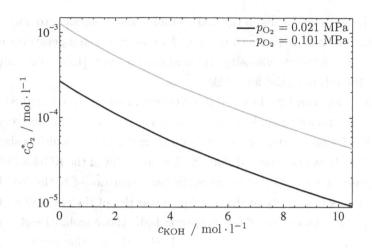

Figure 7.3.: Concentration of dissolved oxygen for two different partial pressures of O_2 and as a function of KOH-molarity; calculated with correlations given by Tromans [63] at $T = 298$ K.

propose an equation to calculate the limiting current density, i_{limit}, for passive operation of a gas diffusion electrode, so that:

$$i_{limit} = \frac{z_e^{air} \cdot \mathbf{F}}{\delta_{film}^{air}} \cdot D_{O_2, KOH} \cdot c_{O_2}^* \qquad (7.7)$$

where z_e^{air} is the number of electrons exchanged in the electrode reaction. The film thickness of liquid electrolyte above the catalyst layer, δ_{film}^{air}, might change with time and battery operation. $D_{O_2, KOH}$ and $c_{O_2}^*$ depend thereby on the electrolyte molarity.

7.4. Temperature

Basically, all parameters for the reaction kinetics of reaction (I) to (V), the transport processes for the reactants and products for the chemical and electrochemical reactions, and consequently all electrode potentials are affected by the temperature in the ZAB: From oxygen solubility, to diffusion coefficients, to reaction rate constants. Moreover, low temperatures can decrease the electrolyte viscosity [76] and conductivity [19], which might decrease the cell potential for ZABs.

Another issue can be addressed if ZABs are intended to be applied for example in automotive applications: The freezing point of the electrolyte, i.e. the KOH-solution, depends strongly on its molarity [89], which is shown in figure 7.4. It can be observed that the freezing point of the KOH-solution is the lowest at molarities of about 8. Below molarities of 8, the freezing point increases drastically so that for low molarities of the electrolyte, the freezing point of pure water 273 K is approached. Above molarities of 8, the freezing point increases slightly but is still below the freezing point of pure water. Since ZABs are normally operated around molarities of 6, they can easily withstand very harsh winter conditions (coldest days with $T = 233$ K) that vehicles are exposed to [90]. By implication, a significant loss or gain of water, and thus an increase or decrease of the electrolyte molarity, might lead to freezing of the electrolyte, which in turn might cause serious performance drops and ZAB damage.

The maximal solubility concentration $c^{sol,max}$ of the salts KOH and K_2CO_3 in water is shown as a function of temperature in figure 7.5. At 298 K the solubility concentration of KOH is 2.5 times greater than for K_2CO_3. This underlines how the low solubility of K_2CO_3 can lead to salt formation in the pores of the air electrode. Both maximal solubility concentrations increase slightly linear with temperature, whereas the slope is steeper for KOH-salts than for K_2CO_3-salts.

All in all, the operating temperature affects almost every aspect of ZAB operation. However, not every ZAB property and its temperature dependency is available in literature. Thus, a temperature analysis is not included for the mathematical modeling in this thesis. For sake of completeness, the impact of temperature on the cell potential and impedance was however briefly investigated for ZAB button cells (see subchapter B.5 in the appendix).

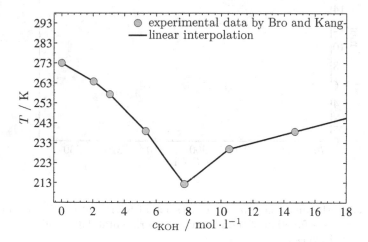

Figure 7.4.: Freezing point of KOH-solution as a function of KOH-molarity; reproduced with experimental data by Bro and Kang [89].

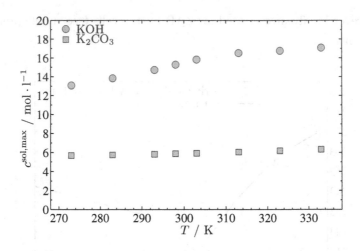

Figure 7.5.: Maximal solubility concentration of KOH and K_2CO_3 in water as a function of temperature; experimental data as given by Haynes [91].

8. Model Approach to Reveal Air-Composition Impact

In this chapter, first a brief overview on existing model approaches to describe ZABs and further electrochemical systems is given. Subsequently, the set of equations for the versatile and expandable basic model for electrically rechargeable ZABs is introduced. The basic model is then modified to account for certain scenarios of air-composition and operation strategies: (a) reference scenario, (b) relative humidity scenario, (c) active operation scenario, (d) carbon dioxide scenario, and (e) oxygen scenario. All presented equations and the main thoughts on the model approach, as well as the underlying assumptions are based on the publication [85].

8.1. State of the Art: Existing Model Approaches

The main justification for the model-based analysis of electrochemical systems arises if one considers the complex or time-consuming experimental methods that are currently discussed in research (see for example DEMS-measurements on lithium-oxygen batteries [92]). These experiments can address all previously mentioned issues for long-term stable operation of ZABs as presented in subchapter 1.2. However, if they would be conducted to cope with all challenges of zinc electrode, air electrode, separator and electrolyte, and air-composition impact, they would imply enormous effort [71]. Furthermore, if experiments are conducted separately, i.e. for each electrode, they do not yield results for the interaction of all occurring reaction and transport processes with each other in the ZAB. Consequently, it would be difficult to draw conclusions from electrode measurements for the full cell ZAB performance. Besides, this would require an even more comprehensive set of measurements, and more advanced experimental set-ups and diagnosis techniques. This especially holds for the incorporation of a reference electrode into electrochemical cells as it might interfere with the processes of interest.

By implication the mathematical description of the diverse reaction and transport processes within ZABs and the subsequent analytical or numerical solution of the equation systems serve as time-efficient and reliable tools to systematically analyze the entire set of processes in the electrochemical system of interest.

The article by Newman and Tobias from 1962 on the *Theoretical Analysis of Current Distribution in Porous Electrodes* [93] was paving the way for the model-based research on electrochemical systems of various kinds and is praised as such in the field of electrochemistry (see [94]). Their approach was successfully applied and adapted to analyze a multitude of battery systems such as lithium-ion batteries ([95] and [96]), and much more recently lithium-air batteries [97]. The standard literature reference for the mathematical description of electrode kinetics, transport processes in porous electrodes, and the numerical solution of coupled ordinary differential equations is another publication by Newman, the book *Electrochemical Systems* [98].

Physical and Chemical Models

Relatively few models exist that describe parts of ZABs or entire ZABs: namely a zinc electrode model [9], a primary ZAB model [99] and a model for an electrically rechargeable ZAB [71]. Furthermore, models of alkaline (see [100] and [101]) or zinc-silver oxide secondary batteries are introduced with the aim to analyze the shape change of the zinc electrode [102]. All models mentioned above, apply more or less the one-dimensional macroscopic approach for porous battery electrodes by Newman and Tobias, and thus consider a system of coupled discretized non-linear differential and algebraic equations. They include the calculation of species concentrations and electrode overpotentials as state variables. Additionally, all aforementioned models are based on the description of chemical reactions and transport processes. Commonly, reaction rate expressions (i.e. Butler-Volmer or Tafel approach), precipitation and solubility equations, and the Nernst-Planck equation, which is applied to describe the species transport in porous electrodes, are used to reflect the physical and chemical phenomena occurring. The usage of the Nernst-Planck equation will be elucidated in more detail in subchapters 8.2 and 9.3.

The research paper by Deiss et al. [71] is an important basis for this thesis. They were the first to introduce a model description of a complete electrically rechargeable ZAB. However, they apply simplified equations for the air electrode. Their focus is primarily on the description of the processes in the zinc electrode for charge and discharge operation. This enables to analyze and optimize ZAB design parameters, such as separator thickness and zinc electrode thickness. However, the influence of the air-composition on the ZAB operation is not considered. A description of a battery with an air electrode open to the environment and the respective impact of air-composition such as relative humidity, oxygen content and carbon dioxide content on battery operation is not included in their work.

Equivalent Circuit Models

To describe electrochemical cells with equivalent circuits is another common modeling method. These circuits are chosen in order to reflect the behavior of the electric and electrochemical parts within the electrochemical cell of interest. With this approach, all physical and chemical processes for an electrochemical system that is subjected to a polarization are assumed to occur according to standard electrical circuit elements such as resistors, capacitors, and their wiring in series or parallel to each other [48].

Often equivalent circuits are used to fit experimental results from electrochemical impedance spectroscopy measurements (see subchapter 3.1) and to determine the parameters for the elements in the electrical equivalent circuit (see for example [103] and [104] for air electrode, and [105] for zinc electrode equivalent circuit models).

Equivalent circuit models are more or less black box models, for which solely the relation of input and output is of interest. They do not embed the description of the interactions between reaction and transport processes inside the electrochemical cell on physical and chemical level. However, they might be more efficient from the computational point of view, since they only require the evaluation of algebraic equations instead of ordinary or partial differential equations.

Further Model Approaches

The existence of a multitude of physical and chemical processes in all kinds of battery types has led to the development of a huge variety of model approaches on different levels of predictability and computational demands [106]. Thereof, first principle approaches [107], and approaches on multiple scales [108] are applied to design and optimize electrochemical systems. Recent advances show that standard continuum methods, molecular simulations, and coarse-grained simulation methods are only a small part of the current research progress (see [108], p. 300-306).

In summary, mathematical modeling and simulation is in particular helpful to analyze the reaction and transport processes within electrochemical cells, and also within ZABs, since these model-based methods can for example reveal detailed information on the electrode potentials or the spacial distribution of species within the cells without complex experimental set-ups, and time-consuming experiments.

8.2. Basic Model

In this subchapter, the versatile and expandable ZAB basic model is derived, which is later modified to account for certain scenarios of air-composition impacts. In general, molar balances for each species in the liquid and solid phase at both electrodes are applied. The model description is given in general terms and is as such not limited to a certain battery geometry. However, it is intended to represent a button cell ZAB geometry similar as depicted in figure 1.1.

Reaction Rates

The reaction rate of the electrochemical conversion of Zn (reaction (I)) is described with a Butler-Volmer approach. The chemical reaction of zincate (reaction (II)) is accounted for with a saturation approach. Both equations are taken from [71] and [9], respectively, so that:

$$
r_{\mathrm{I}} = - \left[k_{\mathrm{I}}^{\mathrm{c}} \cdot \frac{c_{\mathrm{Zn(OH)}_4^2}^{\mathrm{zinc}}}{c^{\mathrm{ref}}} \cdot \exp\left(-\frac{(1 - \alpha^{\mathrm{c,zinc}}) \cdot \mathbf{F}}{\mathbf{R} \cdot T} \cdot \eta^{\mathrm{zinc}} \right) \right.
$$
$$
\left. - k_{\mathrm{I}}^{\mathrm{a}} \cdot \left(\frac{c_{\mathrm{OH}}^{\mathrm{zinc}}}{c^{\mathrm{ref}}} \right)^4 \cdot \frac{n_{\mathrm{Zn}}}{n^{\mathrm{ref}}} \cdot \exp\left(\frac{\alpha^{\mathrm{a,zinc}} \cdot \mathbf{F}}{\mathbf{R} \cdot T} \cdot \eta^{\mathrm{zinc}} \right) \right]
\tag{8.1}
$$

$$
r_{\mathrm{II}} = k_{\mathrm{II}} \cdot \left(c_{\mathrm{Zn(OH)}_4^2}^{\mathrm{zinc}} - c_{\mathrm{Zn(OH)}_4^2}^{\mathrm{sat}} \right)
\tag{8.2}
$$

with k being the reaction rate constant, n the molar amount of a species, c the concentration of a species, α the symmetry coefficient of the electrode reaction, $\mathbf{F}$ Faraday's constant, $\mathbf{R}$ the universal gas constant, and η the electrode overpotential.

The reaction rate of the ORR and OER at the air electrode (reaction (IV)) is described with a Butler-Volmer approach as given by [71]:

$$
r_{IV} = + \left[k_{IV}^a \cdot \frac{c_{H_2O}^{air}}{c^{ref}} \cdot \left(\frac{c_{O_2}^*}{c^{ref}} \right)^{\frac{1}{2}} \cdot \exp \left(-\frac{(1 - \alpha^{c,air}) \cdot \mathbf{F}}{\mathbf{R} \cdot T} \cdot \eta^{air} \right) \right.
$$
$$
\left. - k_{IV}^c \cdot \left(\frac{c_{OH}^{air}}{c^{ref}} \right)^2 \cdot \exp \left(\frac{\alpha^{a,air} \cdot \mathbf{F}}{\mathbf{R} \cdot T} \cdot \eta^{air} \right) \right] \tag{8.3}
$$

where $c_{H_2O}^{air}$ is the concentration of the solvent water in the electrolyte at the air electrode. Oxygen dissolves in the electrolyte, and reacts with water electrochemically to hydroxide ions during discharge, and evolves during the charge of ZABs. Hence, it is accounted for as $c_{O_2}^*$, the concentration of dissolved oxygen in the liquid electrolyte directly at the catalyst layer of the air electrode, in equation (8.3).

For the basic model and all its modifications (except for the oxygen scenario), it is assumed that the amount of reacting oxygen in the air electrode is restored infinitely fast, and no accumulation of dissolved oxygen in the electrolyte is considered. To implement these assumptions, $c_{O_2}^*$ is expressed at all times by its solubility in the KOH-solution. The solubility of oxygen is expressed with an algebraic expression that is given by Tromans [63], and stated in equation (A.31) in the appendix. It is to be noted that $c_{O_2}^*$ is thereby a function of the time dependent KOH-molarity in the air electrode. This enables to investigate the impact of the KOH-molarity and the oxygen solubility on the air electrode's OCP and overpotential. However, it is assumed that $c_{O_2}^*$ does not change significantly with the current density applied, which is justified for small current densities that are also reported for commercially available ZAB button cells.

Molar Balance Equations

The molar concentration for species $k = OH^-$, H_2O and $Zn(OH)_4^{2-}$ in the liquid electrolyte within the electrode j (indicating either zinc or air electrode) is:

$$c_k^j = \frac{n_k^j}{V_{\text{electrolyte}}^j} \tag{8.4}$$

where the electrolyte possesses a volume of $V_{\text{electrolyte}}$. The accumulation of species k in the liquid phase of the ZAB are calculated with the following molar balance in concentration form:

$$\frac{dc_k^j}{dt} = \frac{+J_k^{j,\text{diff}} + J_k^{j,\text{mig}} + J_k^{j,\text{conv}}}{V_{\text{electrolyte}}^j} + \frac{\sum_i \nu_{k,i} \cdot r_i}{V_{\text{electrolyte}}^j} - \frac{c_k^j}{V_{\text{electrolyte}}^j} \cdot \frac{dV_{\text{electrolyte}}^j}{dt} \tag{8.5}$$

with $\nu_{k,i}$ being the stoichiometric coefficient for species k in reaction i. Thereby, homogenous concentrations within each electrode are assumed, similar as in continuous stirred-tank reactors. Consequently, no concentration gradients along the electrode dimensions are included. However, concentration gradients between zinc and air electrode are accounted for with the molar flow rates of exchange, J, in equation (8.5). They are namely considered as diffusion (diff), migration (mig) and convection (conv) molar flow rates. They are applied with the Nernst-Planck equation for diluted solutions in this thesis, and are given in more detail in equations (A.15)-(A.17) in the appendix. For H_2O in the zinc electrode in equation (8.5), the migration molar flow rate of exchange, J^{mig}, is omitted for the basic model because water molecules are considered to possess neutral charge. However, water molecules can move along with the charged ions in the electrolyte. This will be briefly investigated in subchapter 9.3.1.

The accumulation of solid ZnO at the zinc electrode is expressed by the following equation:

$$\frac{dn_{\text{ZnO}}}{dt} = \nu_{\text{ZnO,II}} \cdot r_{\text{II}} \tag{8.6}$$

Since Zn is not allowed to enter or leave the system, nor Zn is allowed to enter the air electrode, the amount of converted solid Zn at the zinc electrode is expressed by an algebraic equation:

$$n_{Zn} = n_{Zn}^{total,t=0} - n_{ZnO} - n_{Zn(OH)_4^2}^{zinc} - n_{Zn(OH)_4^2}^{air} \tag{8.7}$$

For the basic model, no gaseous or liquid water is allowed to be exchanged with the surrounding. Hence, liquid H_2O in the air electrode appears in the basic model in an algebraic equation, and is obtained by balancing the molar amount of species in the ZAB containing H atoms, so that:

$$
\begin{aligned}
n_{H_2O}^{air} = & +\frac{1}{2} \cdot \left(n_H^{total,t=0} - n_{OH}^{air} - 4 \cdot n_{Zn(OH)_4^2}^{air} \right. \\
& \left. -2 \cdot n_{H_2O}^{zinc} - n_{OH}^{zinc} - 4 \cdot n_{Zn(OH)_4^2}^{zinc} \right)
\end{aligned} \tag{8.8}
$$

where $n_H^{total,t=0}$ is the initially present overall amount of hydrogen atoms in the liquid electrolyte of the ZAB. Thereby, the electrolyte volume in the separator is considered to be much smaller than the volume of electrolyte in the zinc electrode and in the air electrode. Thus, the hydrogen atoms in the separator are neglected in equation (8.8).

Volume Change

As explained in subchapter 1.1, and observed experimentally with the X-ray tomography measurements in subchapter 5.2, the solid volume in the zinc electrode changes with time during ZAB operation, which is accounted for with a change of the solid volume inside the zinc electrode, so that:

$$V_{solid}^{zinc} = n_{Zn}(t) \cdot \tilde{V}_{Zn} + n_{ZnO}(t) \cdot \tilde{V}_{ZnO} \tag{8.9}$$

where $\tilde{V}$ is the partial molar volume of the respective species. The change of the solid volume in the zinc electrode with time is expressed by:

$$\frac{dV_{\text{solid}}^{\text{zinc}}}{dt} = \nu_{\text{Zn,I}} \cdot r_{\text{I}} \cdot \tilde{V}_{\text{Zn}} + \nu_{\text{ZnO,II}} \cdot r_{\text{II}} \cdot \tilde{V}_{\text{ZnO}} \tag{8.10}$$

A pressure increase in the zinc electrode and an expansion of the electrode housing is omitted, thus the total electrode volume does not change with time:

$$\frac{dV_{\text{electrode}}^{\text{zinc}}}{dt} = \frac{dV_{\text{solid}}^{\text{zinc}}}{dt} + \frac{dV_{\text{electrolyte}}^{\text{zinc}}}{dt} = 0 \tag{8.11}$$

Thereby, $dV_{\text{electrolyte}}^{\text{zinc}}/dt$ is the change of the liquid electrolyte volume in the zinc electrode due to the change in solid volume, which can be seen as convective flow. Rewriting equation (8.11) yields an expression given by [109], so that the resulting convective flow of electrolyte, F^{conv}, is expressed as:

$$F^{\text{conv}} = -r_{\text{I}} \cdot \sum_i \nu_{i,\text{I}} \cdot \tilde{V}_i$$

$$-r_{\text{II}} \cdot \sum_k \nu_{k,\text{II}} \cdot \tilde{V}_k$$

$$+\sum_g J_g^{\text{zinc,diff}} \cdot \tilde{V}_g$$

$$+\sum_g J_g^{\text{zinc,conv}} \cdot \tilde{V}_g$$

$$+\sum_h J_h^{\text{zinc,mig}} \cdot \tilde{V}_h \tag{8.12}$$

with $i = \text{Zn, OH}^-, \text{ZnOH}_4^{2-}$, $k = \text{ZnO, OH}^-, \text{ZnOH}_4^{2-}, \text{H}_2\text{O}$, $g = \text{OH}^-$, $\text{K}^+, \text{ZnOH}_4^{2-}, \text{H}_2\text{O}$ and $h = \text{OH}^-, \text{K}^+, \text{ZnOH}_4^{2-}$.

In the air electrode, the convective flow rate is defined as coming from the zinc electrode and is therefore denoted with a negative sign (see equation (A.20) in the appendix). Consequently, the volume change of electrolyte in the air electrode becomes:

$$\frac{\mathrm{d}V_{\mathrm{electrolyte}}^{\mathrm{air}}}{\mathrm{d}t} = -F^{\mathrm{conv}} + r_{\mathrm{IV}} \cdot \sum_m \nu_{m,\mathrm{IV}} \cdot \tilde{V}_m$$
$$+ \sum_g J_g^{\mathrm{air,diff}} \cdot \tilde{V}_g + \sum_g J_g^{\mathrm{air,conv}} \cdot \tilde{V}_g + \sum_h J_h^{\mathrm{air,mig}} \cdot \tilde{V}_h \tag{8.13}$$

with $m = \mathrm{OH}^-$ and H_2O. It is to be noted that the partial molar volume of oxygen is comparably small and neglected in equation (8.13), as suggested by Kimble et al. [110].

Charge Balances

Each battery electrode is additionally described with an ordinary differential equation (ODE) to account for transient changes of the respective electrode overpotential η:

$$\frac{\mathrm{d}\eta^{\mathrm{zinc}}}{\mathrm{d}t} \cdot C_{\mathrm{DL}}^{\mathrm{zinc}} = i^{\mathrm{cell}} - \frac{z_e^{\mathrm{zinc}} \cdot \mathbf{F} \cdot r_{\mathrm{I}}}{A_{\mathrm{electrode}}^{\mathrm{zinc}}} \tag{8.14}$$

$$\frac{\mathrm{d}\eta^{\mathrm{air}}}{\mathrm{d}t} \cdot C_{\mathrm{DL}}^{\mathrm{air}} = -i^{\mathrm{cell}} + \frac{z_e^{\mathrm{air}} \cdot \mathbf{F} \cdot r_{\mathrm{IV}}}{A_{\mathrm{electrode}}^{\mathrm{air}}} \tag{8.15}$$

with z_e^j being the number of electrons participating in the electrochemical reaction at electrode j, i. e. $= 2$.

Electroneutrality

During ZAB operation, electroneutrality is assumed to hold in the electrolyte. Thus, an ionic species charge balance at each electrode is applied as suggested by [98]:

$$z_{\mathrm{K}^+} \cdot c_{\mathrm{K}^+}^{\mathrm{zinc}} + z_{\mathrm{Zn(OH)}_4^2} \cdot c_{\mathrm{Zn(OH)}_4^2}^{\mathrm{zinc}} + z_{\mathrm{OH}} \cdot c_{\mathrm{OH}}^{\mathrm{zinc}} = 0 \tag{8.16}$$

$$z_{K+} \cdot c_{K+}^{air} + z_{Zn(OH)_4^2} \cdot c_{Zn(OH)_4^2}^{air} + z_{OH} \cdot c_{OH}^{air} = 0 \qquad (8.17)$$

with z_k being the charge number for the ionic species k.

Cell Potential

The ZAB cell potential is given by the open circuit potential, the over-potentials at zinc and air electrode and ohmic losses. It is applied as follows:

$$E^{cell} = E^{0,cell} - \eta^{zinc} + \eta^{air} - \eta_{ionic}^{sep} - \eta_{ohmic}^{zinc} \qquad (8.18)$$

where $E^{0,cell}$ is given by the Nernst potentials for the electrodes, represented by equation (A.28) and (A.29) in the appendix. Ohmic potential losses are solely accounted for as a result of ion transport resistance in the separator (possessing a thickness of δ^{sep}) with η_{ionic}^{sep}, and at the zinc electrode with η_{ohmic}^{zinc}. The potential losses are applied as:

$$\eta_{ohmic}^{zinc} = \frac{i^{cell} \cdot A_{electrode}^{zinc}}{\sigma^{zinc} \cdot \delta^{zinc}} \qquad (8.19)$$

$$\eta_{ionic}^{sep} = - \Delta\Phi(i^{cell}) \cdot \delta^{sep} \qquad (8.20)$$

$\Delta\Phi$ in equation (8.20) thereby represents the potential gradient between the electrolyte phase of air electrode and zinc electrode. It is given in more detail in equation (A.26) in the appendix. σ^{zinc} in equation (8.19) is the overall zinc electrode conductivity, and δ^{zinc} is the zinc electrode thickness. They are expressed with the following equations, so that:

$$\sigma^{zinc} = x_{Zn} \cdot \sigma^{Zn} + x_{ZnO} \cdot \sigma^{ZnO} \qquad (8.21)$$

$$x_{Zn} = \frac{n_{Zn}}{n_{Zn} + n_{ZnO}} \qquad (8.22)$$

$$x_{\text{ZnO}} = \frac{n_{\text{ZnO}}}{n_{\text{Zn}} + n_{\text{ZnO}}} \tag{8.23}$$

with x being the molar fraction of the solid species Zn and ZnO.

8.3. Scenarios to Account for Air-Composition Impact

To reveal the limitations due to the impact of the air-composition on the long-term and stable operation of ZABs, the methodology of scenario-based modeling (see [111]) is applied in the following. Thereby, the impact factors introduced in chapter 7 are included in separate scenarios with selected sets of equations, to independently address the underlying physical and chemical phenomena of interest for the respective scenario.

(a) Reference Scenario

The basic model introduced in subchapter 8.2 is set up as the reference scenario. It does not account for any water loss or gain or side reactions in the air electrode. The supply of oxygen is at all times sufficient and is not limiting at elevated current densities. Simulation results that are obtained with this scenario are intended to show a stable reference case for the cycling behavior of ZABs.

(b) Relative Humidity Scenario

The relative humidity scenario is a modification of the reference scenario that focuses on the exchange of water vapor in the air electrode with the surrounding during passive ZAB operation. None of the other aforementioned air-composition influences is considered for this scenario, and all equations of the reference scenario hold unless indicated otherwise. A schematic of the air electrode considered for this scenario, and the occurring exchange with the surrounding air is depicted in figure 8.1. The molar fraction of gaseous water in the environmental air, $y_{H_2O(g)}^{env}$, is set according to the environmental temperature T^{env} and the environmental relative humidity RH^{env}. The equilibrium molar fraction of gaseous water directly above the liquid electrolyte is denoted as $y_{H_2O(g)}^{sat}$, i.e. water saturated gas is assumed. Since liquid KOH-electrolyte is present, the molar fraction of water above the electrolyte, $y_{H_2O(g)}^{sat}$, depends on the battery temperature and the liquid electrolyte concentration. If during battery operation the molar fraction of gaseous water at the electrode surface differs from the one in the surrounding, a molar flow rate , $J_{H_2O(g)}$, of gaseous water through the GDL is evoked. This molar flow rate through the GDL is described by Fick's law, depends on the diffusion coefficient of gaseous water in the GDL, $D_{H_2O(g)}^{eff,GDL}$, and the GDL properties, such as the thickness δ^{GDL} and the porosity ε^{GDL}.

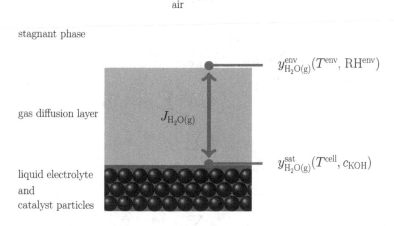

Figure 8.1.: Schematic of the exchange of gaseous water in the air electrode with surrounding air considered for the relative humidity scenario. Reprinted from [85] with permission from Elsevier.

On this basis, the introduced basic model is modified as follows. First, the algebraic equation for the water amount in the air electrode (see equation (8.8)) is replaced by an ODE for the accumulation of liquid water in the air electrode because the total molar amount of water in the entire ZAB is not constant during operation for this scenario. Second, the electrolyte ODE (see equation (8.13)) is extended by $J_{H_2O(g)}$, the molar flow rate of gaseous water exchange. This yields:

$$
\begin{aligned}
\frac{dc_{H_2O}^{air}}{dt} &= \frac{+J_{H_2O}^{diff} + J_{H_2O}^{conv}}{V_{electrolyte}^{air}} \\
&+ \frac{\nu_{H_2O,IV} \cdot r_{IV}}{V_{electrolyte}^{air}} \\
&+ J_{H_2O(g)} \\
&- \frac{c_{H_2O}^{air}}{V_{electrolyte}^{air}} \cdot \frac{dV_{electrolyte}^{air}}{dt}
\end{aligned}
\tag{8.24}
$$

and:

$$
\begin{aligned}
\frac{dV_{electrolyte}^{air}}{dt} &= -F^{conv} \\
&+ r_{IV} \cdot \sum_m \nu_{m,IV} \cdot \tilde{V}_m \\
&+ \sum_g J_g^{air,diff} \cdot \tilde{V}_g \\
&+ \sum_g J_g^{air,conv} \cdot \tilde{V}_g \\
&+ \sum_h J_h^{air,mig} \cdot \tilde{V}_h \\
&+ J_{H_2O(g)} \cdot \tilde{V}_{H_2O}
\end{aligned}
\tag{8.25}
$$

$J_{H_2O(g)}$ is introduced as:

$$J_{H_2O(g)} = - \frac{p_{atm}}{R \cdot T} \cdot D_{H_2O(g)}^{eff,GDL} \cdot \Delta y \cdot \frac{A^{GDL}}{\delta^{GDL}} \qquad (8.26)$$

with A^{GDL} being the effective area for exchange with the environment. The driving force for the molar flow is thereby the difference of the molar fractions of gaseous water above the liquid electrolyte and in the surrounding air, Δy. It is expressed with a correction for back-diffusion during the flow and evaporation of gaseous water (Stefan-flow), so that:

$$\Delta y = \frac{y_{H_2O(g)}^{sat} - y_{H_2O(g)}^{env}}{1 - y_{H_2O(g)}^{sat}} \qquad (8.27)$$

$y_{H_2O(g)}^{sat}$-values for this work are calculated with electrolyte concentration and temperature dependent correlations given by Balej [80]. $y_{H_2O(g)}^{env}$ is expressed as follows:

$$y_{H_2O(g)}^{env} = RH^{env} \cdot \frac{p_{H_2O(g)}(T^{env})}{p_{atm}} \qquad (8.28)$$

A derivation for this equation is given in subchapter 7.1. $p_{H_2O(g)}$-values for this work are calculated with temperature dependent correlations given by Stull [81], as given in the appendix in equation (A.47).

(c) Active Operation Scenario

If ZABs are meant for the application in automotive applications or for stationary power supply, larger current densities and/or ZAB stacks with a multitude of cells, similar to fuel cell applications, might be applied. Consequently, the supplied oxygen should be fed with excess via a large flow rate of pure oxygen or air to ensure that the oxygen concentration at the CL is sufficiently large enough to maintain the electrochemical reaction (IV), and thus the current density desired. A high flow rate of humidified or dry gas, either pure oxygen or air, will take up or bring in water into the air

electrode of the ZAB and can possibly contribute to an early depletion or
high accumulation of water in the air electrode.

The active operation scenario is a modification of the relative humidity
scenario to account for the exchange of water vapor during the active supply
of air or oxygen. A schematic of the air electrode with the inlet and outlet
flow of the oxygen or air supplied is given in figure 8.2. For this scenario, a
gas flow that contains a molar fraction of gaseous water $y_{H_2O(g)}^{env}$ (as a function
of environmental temperature T^{env} and environmental relative humidity
RH^{env}) is entering the air electrode at a molar flow rate of J_{gas}, is then
instantaneously uniformly mixed with the gaseous water above the liquid
electrolyte in the GDL, (similar as in a continuous stirred-tank reactor), and
thus exits the air electrode with $y_{H_2O(g)}^{sat}(T^{env})$, the molar fraction of gaseous
water directly above the liquid electrolyte. As a consequence, gaseous water
will be added to the liquid electrolyte in the air electrode or will be taken
from the liquid electrolyte, respectively, with a molar flow rate $J_{H_2O(g)}$.

The inlet molar flow is multiplied by an oxygen excess factor, λ_{O_2}, which
is usually applied in fuel cell applications to ensure a sufficient amount of
reactant at the reaction zone to avoid oxygen starvation. λ_{O_2} is coupled
with Faraday's law to calculate the needed molar flow of supplied gas that
contains oxygen, so that:

$$J_{gas} = \lambda_{O_2} \cdot J_{O_2}^{stoic} = \lambda_{O_2} \cdot \frac{i^{cell} \cdot A_{electrode}^{air}}{F \cdot z^{air}} \cdot \frac{1}{y_{O_2}} \qquad (8.29)$$

In this definition, the value for λ_{O_2} is fixed so that the molar flow rate
depends on the current density applied. J_{gas} is set to zero in the case of
battery charge because no oxygen supply would be needed for the OER at
the air electrode. If the supplied gas is air, the factor $1/y_{O_2}$ in equation (8.29)
becomes $1/0.21$ to account for the low amount of 21 vol% of O_2 in air, whereas
$1/y_{O_2} = 1$ for the operation with pure oxygen. The molar flow rate of gaseous
water, which contributes to the accumulation of water in the air electrode,

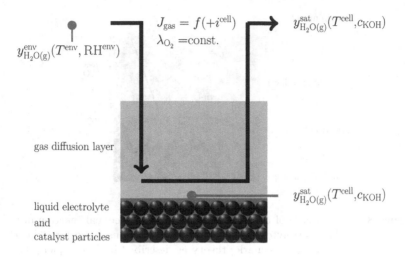

Figure 8.2.: Schematic for the active operation scenario; air electrode indicated with inlet and outlet for the gas flow.

as described in equation (8.25), is then expressed as:

$$J_{H_2O(g)} = \Delta y \cdot J_{gas} \qquad (8.30)$$

where Δy is as described in equation (8.27). It can be seen that $J_{H_2O(g)}$ strongly depends on the supplied flow rate of gas but as well on the water vapor equilibrium between environment and electrolyte, which is given by the molarity of the applied electrolyte solution.

(d) Carbon Dioxide Scenario

The carbon dioxide scenario is another modification of the basic model to solely account for the influence of carbon dioxide on passive ZAB operation. Influences of relative humidity and oxygen from the surrounding are not accounted for. The uptake of CO_2 is described as diffusion and absorption process, which is given by the geometry and structure of the air electrode. A schematic of the air electrode with CO_2-absorption considered for this

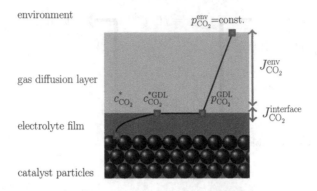

Figure 8.3.: Schematic of the CO_2-absorption with chemical reaction in
the electrolyte film in the air electrode considered for the carbon dioxide scenario; theory as described by Levenspiel [112].
Reprinted from [85] with permission from Elsevier.

scenario is shown in figure 8.3. In presence of absorbed CO_2, the species
CO_3^{2-} will form in the liquid electrolyte. Therefore, the basic model
equations are adapted for this scenario as follows. First, two additional
balance equations for CO_3^{2-} at zinc and air electrode are included according
to equation (8.5). Second, the carbonation of hydroxide ions is applied
as introduced by Pohorecki and Moniuk [87] with the reaction rate r_V, as
given in subchapter 7.2 in equation (7.6). The concentration of dissolved
carbon dioxide is calculated by a molar balance equation in concentration
form, so that:

$$\frac{dc_{CO_2}^*}{dt} = \frac{\nu_{CO_2,V} \cdot r_V + J_{CO_2}^{\text{interface}}}{V_{\text{electrolyte}}^{\text{air}}}$$
$$- \frac{c_{CO_2}^*}{V_{\text{electrolyte}}^{\text{air}}} \cdot \frac{dV_{\text{electrolyte}}^{\text{air}}}{dt} \qquad (8.31)$$

where $J_{CO_2}^{interface}$ is the molar flow rate of CO_2 coming from the GDL that will be absorbed in the liquid electrolyte of the air electrode. The basis for this approach is the theory for gas absorption with reaction by Levenspiel [112]. Carbon dioxide is absorbed in the liquid electrolyte film above the catalyst particles, from where it is transported to the bulk of the liquid electrolyte, while undergoing the carbonation reaction. Thereby equilibrium is established at the interface of GDL and gas phase, resulting in the concentration $c_{CO_2}^{*GDL}$.

$J_{CO_2}^{interface}$ is described as found in [82]:

$$J_{CO_2}^{interface} = \left(\frac{p_{CO_2}^{GDL} \cdot En}{H_{CO_2}} - c_{CO_2}^{*GDL} \right) \cdot k_{CO_2}^{electrolyte} \cdot A^{GDL} \qquad (8.32)$$

hereby the Hatta modulus, Ha, will determine the enhancement factor, En, for the absorption process with reaction in the liquid phase. In this thesis, Ha and En are calculated as given by Kucka et al. [82], applying the mass transfer coefficient $k_{CO_2}^{electrolyte}$. The equations for Ha and En are stated in the appendix in equations (A.32) and (A.33).

The concentration of dissolved carbon dioxide directly at the interface of GDL and liquid electrolyte film, $c_{CO_2}^{*GDL}$, is influenced by the ionic strength of the electrolyte ($S_{ionic} = 1/2 \cdot \sum_k z_k^2 \cdot c_k$, [98]), and the carbon dioxide partial pressure in the GDL ($p_{CO_2}^{GDL}$), which is expressed with Henry's law constant, H_{CO_2}, as given in the appendix in equation (A.34). $p_{CO_2}^{GDL}$ is calculated by Fick's diffusion assuming a linear profile from the constant partial pressure of carbon dioxide in the surrounding, $p_{CO_2}^{env}$, through the GDL (see equation (A.14) in the appendix).

Finally, equation (8.13), is extended by the volume change in the air electrode due to the carbonation reaction (V), so that:

$$\frac{\mathrm{d}V_{\text{electrolyte}}^{\text{air}}}{\mathrm{d}t} = - F^{\text{conv}}$$

$$+ r_{\text{IV}} \cdot \sum_m \nu_{m,\text{IV}} \cdot \tilde{V}_m$$

$$+ r_{\text{V}} \cdot \sum_o \nu_{o,\text{V}} \cdot \tilde{V}_o$$

$$+ \sum_g J_g^{\text{air,diff}} \cdot \tilde{V}_g$$

$$+ \sum_g J_g^{\text{air,conv}} \cdot \tilde{V}_g$$

$$+ \sum_h J_h^{\text{air,mig}} \cdot \tilde{V}_h \qquad (8.33)$$

where $o = \text{OH}^-$, H_2O and CO_3^{2-}. h and g are used as in equation (8.12) but are extended by CO_3^{2-}, respectively.

Furthermore, the electroneutrality conditions are modified for this scenario to account for the carbonate ions in the zinc and air electrode, so that:

$$z_{\text{K}^+} \cdot c_{\text{K}^+}^{\text{zinc}} + z_{\text{Zn(OH)}_4^2} \cdot c_{\text{Zn(OH)}_4^2}^{\text{zinc}}$$
$$+ z_{\text{OH}} \cdot c_{\text{OH}}^{\text{zinc}} + z_{\text{CO}_3^2} \cdot c_{\text{CO}_3^2}^{\text{zinc}} \qquad (8.34)$$
$$= 0$$

$$z_{\text{K}^+} \cdot c_{\text{K}^+}^{\text{air}} + z_{\text{Zn(OH)}_4^2} \cdot c_{\text{Zn(OH)}_4^2}^{\text{air}}$$
$$+ z_{\text{OH}} \cdot c_{\text{OH}}^{\text{air}} + z_{\text{CO}_3^2} \cdot c_{\text{CO}_3^2}^{\text{air}} \qquad (8.35)$$
$$= 0$$

(e) Oxygen Scenario

The oxygen scenario is another modification of the basic model to account for the limited oxygen solubility during passive ZAB operation. The influence of the relative humidity and the carbon dioxide is not accounted for in this scenario. The uptake of O_2 is described as diffusion and absorption process without reaction, which is determined by the geometry of the air electrode. A schematic of the air electrode with O_2-absorption considered in this scenario is shown in figure 8.4. Again the absorption theory of Levenspiel [112] is applied. Here, O_2 is solely absorbed and not consumed in the electrolyte film during the transport through it.

Figure 8.4 also illustrates the occurring molar flows of oxygen in the air electrode, and the applied concentration and pressure values for this model. In this scenario, the concentration of dissolved oxygen in the air electrode, $c_{O_2}^*$, changes with time, so that:

$$
\frac{dc_{O_2}^*}{dt} = \frac{\nu_{O_2}\text{IV} \cdot r_{,\text{IV}} + J_{O_2}^{\text{bulk}}}{V_{\text{electrolyte}}^{\text{air}}} \\
- \frac{c_{O_2}^*}{V_{\text{electrolyte}}^{\text{air}}} \cdot \frac{dV_{\text{electrolyte}}^{\text{air}}}{dt}
\tag{8.36}
$$

$J_{O_2}^{\text{bulk}}$ is expressed by balancing the molar flows, and concentration gradients indicated in figure 8.4 so that:

$$
J_{O_2}^{\text{bulk}} = - D_{O_2,\text{KOH}}^{\text{eff}} \cdot \frac{c_{O_2}^* - c_{O_2}^{*\text{GDL}}}{\delta_{\text{film}}^{\text{air}}} \cdot A^{\text{GDL}}
\tag{8.37}
$$

$c_{O_2}^{*\text{GDL}}$ depends on the KOH-molarity and the oxygen partial pressure in the GDL, $p_{O_2}^{\text{GDL}}$. A Henry-type relation [63], given in the appendix in equation (A.31), is used in this chapter of the thesis to calculate $c_{O_2}^{*\text{GDL}}$. $p_{O_2}^{\text{GDL}}$ is calculated by Fick's diffusion assuming a linear profile from the constant $p_{O_2}^{\text{env}}$ through the GDL (see equation (A.14) in the appendix).

The experimental results in subchapter 5.3.1 have shown that the volume of liquid in the GDL, and thus the thickness of the electrolyte film, $\delta_{\text{film}}^{\text{air}}$, above the catalyst layer, could change with operation time. However, $\delta_{\text{film}}^{\text{air}}$ is considered constant for the simulations in this chapter to single out the impact of the oxygen concentration on the ZAB operation.

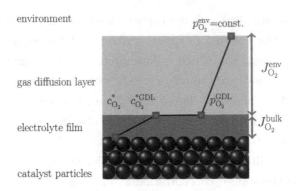

Figure 8.4.: Schematic of the O_2-absorption without chemical reaction in the air electrode considered for the oxygen scenario; theory as described by Levenspiel [112]. Reprinted from [85] with permission from Elsevier.

8.4. Design Parameters and Material Properties

Each of the scenarios introduced requires a set of parameters, such as geometry parameters, material properties and environmental parameters. The parameters applied for the simulations in this chapter of this thesis are given in table A.3 in the appendix.

To ensure reliable electrochemical backing for the simulated cell potential and electrode potentials, the reaction rate constants for the electrochemical reactions at the zinc and air electrode (k_I^a, k_I^c, k_{IV}^a, and k_{IV}^c, indicated with $*$ in table A.3) are obtained by fitting from experimental data as follows. First, the cell potential and the overpotentials are simulated for the reference scenario for polarization curve simulations with a holding time of 60 seconds for an each time new initialized battery at 50% SOD for a randomly chosen set of reaction rate constants. Subsequently, the reaction rate constants are varied by a non-linear least-square fitting method to minimize the difference

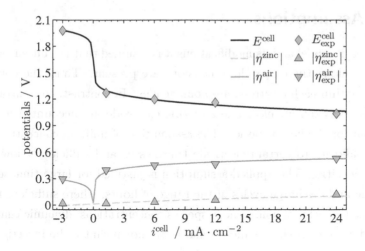

Figure 8.5.: Comparison of the reference scenario simulation results for cell
potential and absolute overpotentials at zinc and air electrode
with experimental data by [71]. The simulated values are obtained with polarization curve simulations with a holding time
of 60 seconds. Each simulated data point is obtained for an each
time new initialized battery at 50% SOD. The initial values are
given in table A.3.

of simulated and experimental data by Deiss et al. [71] for cell potential,
zinc electrode and air electrode overpotential.

A comparative plot of the experimental and the simulated data is depicted
in figure 8.5. The experimentally obtained potentials and the reference
scenario results are in line within a margin of a few mV. This gives a
reliable basis for the simulation results presented later on regarding the
ZAB cell potential and the overpotentials. The currently reported enhanced
overpotentials due to new catalysts at the air electrode (see [21]) could be
used as basis for the fitting of the reaction rate constants for future work.

8.5. Assumptions

For the basic model and its modifications it is assumed that no concentration gradients along the electrode dimensions are present. Thus, the concentrations within each electrode are homogenous. In contrast, concentration gradients between air electrode and zinc electrode are accounted for, in particular for the liquid species. The assumption of uniform concentrations is only valid up to certain electrode thicknesses and sufficiently low cell current densities. The applied assumption is justified for large time scales of charge and discharge cycles in the range of hours, where sufficient time is provided to achieve homogenous species concentrations; dynamic changes in i^{cell} during ZAB charge and discharge are not meant to be investigated with the model approach presented in this chapter.

In addition to the aforementioned assumptions, the following are assumed to hold: The battery is considered to be operated constantly at $T = 298.15$ K and $T^{\text{env}} = 298.15$ K, and all temperature dependent parameters are calculated or taken from literature at this temperature.

The active areas for the electrochemical reactions at the electrodes are considered constant during the entire operation. The thickness of the zinc electrode is also considered constant. The saturation concentration $c^{\text{sat}}_{\text{Zn(OH)}_4^2}$ in the liquid electrolyte is set constant and is independent from the KOH-concentration.

The diffusion coefficients for O_2 and OH⁻ in the aqueous KOH-solution are implemented as independent from the KOH-molarity in this chapter. This is justified if the KOH-concentrations obtained do not vary more than between 4 and 8 mol $\cdot$ l^{-1}.

The volume of electrolyte in the separator is considered to be much smaller than the volume of electrolyte in the zinc electrode. Consequently, the volume of the electrolyte in the separator, and all species in the separator, are neglected in the balance equations.

Furthermore, it is assumed that a change of the liquid electrolyte volume in the pores of the GDL does not significantly increase the amount of water in the pores; the GDL pores can not be flooded with liquid and the liquid film thickness δ_{film}^{air} is constant during the entire operation. As a consequence, the effective diffusion coefficient in the GDL is constant. The flooding and increase in liquid film thickness is however accounted for with the detailed one-dimensional air electrode model presented in chapter 6.

Ideal gas behavior for the exchange of $H_2O(g)$, CO_2 and O_2 is assumed in the respective scenarios. Moreover, the diffusion coefficients are considered for single component, i. e. Fickian, diffusion in N_2. All gaseous species are only accounted for at the air electrode and do not permeate through the separator into the zinc electrode. Dissolved CO_2 and O_2 are not considered to pass through the separator.

9. Simulation Results and Discussions for Air-Composition Impact

In the following chapter, the simulation results for the reference scenario and the scenarios for the air-composition impacts are presented. The respective subchapters aim to assess the challenges that arise from the fact that ZABs are half-open to the surrounding air. The model approach and the simulations conducted in this chapter are based on a previously published journal article [85], and are extended by several simulations.

All equations and scenarios were implemented in Matlab, and the respective set of equations was solved with the ode15s ODE solver. For all simulations, the following initial electrolyte composition is applied, unless indicated otherwise. The electrolyte possesses $6.00 \text{ mol} \cdot \text{l}^{-1}$ K^+. The concentration of $Zn(OH)_4^{2-}$ is set constant to $0.66 \text{ mol} \cdot \text{l}^{-1}$, so that the concentration of OH^- is equal to $4.68 \text{ mol} \cdot \text{l}^{-1}$ due to the electroneutrality in the electrolyte solution, unless indicated otherwise. All other initial conditions at $t = 0$ seconds are given in table A.2.

The majority of the simulation results is obtained for constant current density charge and discharge cycles with 6 hours of charge at $i^{\text{cell}} = -9 \text{ mA} \cdot \text{cm}^{-2}$ and 3 hours of discharge at $i^{\text{cell}} = 18 \text{ mA} \cdot \text{cm}^{-2}$, unless indicated otherwise. The set current density for three cycles is shown in the appendix in figure A.3. With the given initial conditions, one cycle causes a maximal change in SOD of about 17%. It is to be noted that the shown cycle numbers are indicated as discrete values, however the x-axis is to be seen as indicator for continuous time values.

The parameters of the introduced simulations can be adjusted easily, so that for example other current densities can be set as model input. Expanding the simulations presented by various current densities as input value could reveal an optimal operation strategy regarding charge and discharge, which is out of the focus for this thesis.

9.1. Ideal Case Operation of Zinc Air Batteries

As presented in subchapter 1.1, the electrochemical reactions and the ion transport in ZABs are the processes that are fundamental for the working principle of ZABs. If these processes can be maintained and are not disturbed by side reactions due the air-composition impact, e.g. the carbonation reaction, the operation would last for an infinite number of charge and discharge cycles, which is referred to as ideal case operation in the following. This thought is addressed with the reference scenario in this thesis, and will be evaluated and discussed in the following.

To observe the reaction and transport processes during the ideal case operation, the cell potential and overpotentials, the water content in each electrode, and the zinc electrode porosity will be analyzed. For better visualization, the dimensionless relative water content χ is introduced. It is defined in this thesis as the ratio between the actual molar amount of water, variable with time, and the initial molar amount of water at $t = 0$ seconds, so that $\chi = n_{H_2O}(t)/n_{H_2O}^{t=0}$.

Evaluation of the Reference Scenario (a)

The simulation results for the cell potential and the relative water content in zinc and air electrode with increasing cycle number are shown in figure 9.1 (a) and 9.1 (b). The cell potential for charge is approximately 2.10 V and the cell potential for discharge is approximately 1.10 V, which is in line with simulation results by Deiss et al. [71] for similar current densities. In addition, the cell potential observed in figure 9.1 (a) is very flat during charge and discharge, which can also be observed for commercially available ZABs (see [1], p. 13.7).

The SOD changes approximately by 17% for the cycle applied, which is associated with a maximal change in water level in both electrodes of about the same percentage. Since no water gain or loss is accounted for in this scenario, the cell potential and the relative water content in zinc and air electrode progress periodically and reach the same values during each cycle. The operation could continue for an infinite number of charge and discharge cycles. However, recent experimental results on ZAB cycling show that presently only 15 cycles (with full discharge capacity) [37], and 60 cycles (with pulse cycling between 0% and 10% SOD) are achievable [21].

The fact that the water level in figure 9.1 (b) is changing with time at both electrodes, can be explained by the volume expansion of the zinc electrode, as mentioned in subchapter 1.1 and observed experimentally with the X-ray tomography of the zinc electrode in subchapter 5.2. The volume expansion of the solid material during discharge is accompanied by a convective flux of electrolyte out of the zinc electrode, decreasing the volume of liquid electrolyte in the zinc electrode $V_{electrolyte}^{zinc}$ and consequently the relative water content χ^{zinc}. As a result, the relative water content in the air electrode χ^{air} increases. The behavior during charge is then vice versa.

Moreover, the data for the simulation results of the reference scenario reveals that a zinc electrode porosity, ε^{zinc}, at the end of each charge cycle of 0.6 is obtained. This concurs with experimental and model-based findings for similar initial conditions by Mao and White [99]. Together with the X-ray analysis of the volume expansion in the zinc electrode of the in-house set-up (see figure 5.10), this simulation result implies the following: If the entire space between the particles in the zinc electrode is occupied by the formed ZnO during discharge, no void space for the liquid electrolyte would be available anymore in the zinc electrode. This would stop the electro-

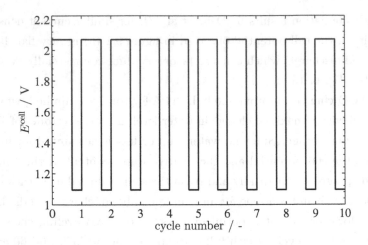

(a)

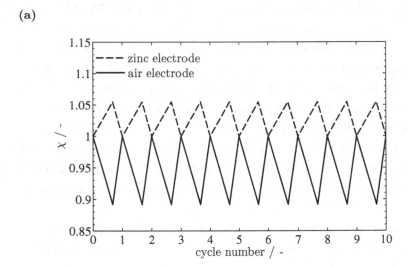

(b)

Figure 9.1.: Simulation results for the reference scenario: (a) cell potential E^{cell}; (b) relative water content χ at the air and zinc electrode.

chemical reaction (I) because the reactant OH^- is not available anymore, and the product $ZnOH_4^{2-}$ might not dissolve in the electrolyte due to a lack of solvent, respectively electrolyte. In conclusion, it is apparent that a carefully planned charge and discharge strategy, and the choice of zinc electrode porosity at the beginning of the operation are crucial for long-term stable ZAB operation.

To resemble the ZA13-type button cell investigated with the X-ray tomography measurements (see subchapter 5.2.3), geometry parameters and initial conditions were set as given in table A.6. Simulations were conducted with these parameters for the reference scenario to extend the experimental results obtained, and to gain further understanding of the electrochemical behavior and the zinc electrode porosity of button cell ZABs during discharge. In particular, discharge curves will be analyzed in the following.

Figure 9.2 (a) illustrates the simulated transient response of the cell potential, i.e. the discharge curve, for various discharge current densities. In general, the cell potential obtained is flat and decreases rapidly to values below 1 V at the end of discharge. For the discharge with $5.42 \, mA \cdot cm^{-2}$ (light gray colored line), the cell potential stays constant with discharge time until a maximal discharge time of 210 hours is reached. Smaller or larger applied current densities extend or diminish the maximal discharge time, respectively. However, the maximum of the SOD is the same for all applied current densities because the model approach accounts for a rather idealized discharge process where the entire amount of zinc particles is accessible until the end of the discharge. For real ZAB discharge, the SOD, and thus the maximal battery capacity that can be withdrawn, strongly depends on the set current density. This is evident for the in this thesis experimentally obtained discharge curves shown in figure 5.5 in subchapter 5.1.

A very steep decay of the cell potential at the end of discharge, as apparent in the simulation results, can not be observed in experimental discharge curves. For these simulations with the reference scenario, the sharp decay of the cell potential is, as implemented, due to the shortage of the molar amount of Zn in the zinc electrode at the end of discharge, which reduces also the zinc electrode conductivity (see equation (8.19)) significantly. In practical ZAB operation, η^{zinc} might be affected by the distribution and morphology of the Zn particles as well. Some zinc particles might be insulated from ionic and electric conductivity at SOD values close to 100%, as shown in the X-ray tomography measurements (see figure 5.9), which could decrease the cell potential at the end of discharge for elevated current densities in a different way than simulated. Moreover, the flooding of the air electrode with liquid electrolyte due to the volume expansion of the zinc electrode (see subchapter 5.3.1) and its impact on the cell potential is not accounted for. The here presented model description could be extended by these effects. However, the analysis in this chapter is aiming on a more detailed description of the impact of the surrounding on ZAB operation, where it might not be crucial to include for example a space and time-dependent model description of the zinc particle conversion and of the liquid electrolyte in the air electrode.

Figure 9.2 (b) depicts the porosity in the zinc electrode during discharge for a zinc electrode porosity at the start of the simulation of $\varepsilon^{zinc} = 0.5$ for the same ZA13 button cell simulation.

Simulation results are again shown for various current densities. It can be observed that at the end of discharge all porosity-values simulated reach 0.25, implying a complete conversion of Zn to ZnO.

In order to validate the cell potentials obtained with the reference scenario, polarization curves are simulated and compared with experimentally obtained polarization curves. Figure 8.5 in subchapter 8.4 depicts the results obtained for the reference scenario cell potential E^{cell} and the absolute overpotentials at zinc and air electrode, compared to experimental data by Deiss et al. [71] for relatively low current densities. It can be observed that the curves obtained are in line with the theoretical polarization behavior of ZABs (see chapter 1.1).

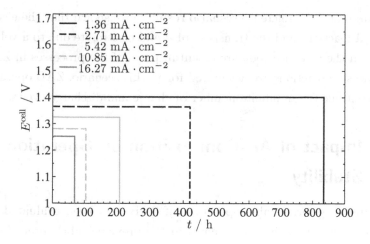

(a)

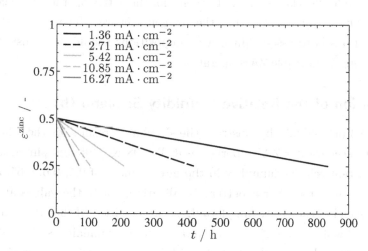

(b)

Figure 9.2.: Simulation results for the discharge at various discharge current densities with parameters of a ZA13-type button cell for the reference scenario: (a) cell potential E^{cell}; (b) porosity ε^{zinc} of the zinc electrode.

In summary, the reference scenario results have shown that the electro-chemical reactions and the transport of liquid electrolyte due to a volume change in the zinc electrode are essential parts of the processes in ZABs. Since no side reactions are accounted for in this scenario, ZAB operation could continue for an infinite number of charge and discharge cycles.

9.2. Impact of Air-Composition on Operation Stability

Ideal and long-term stable operation of ZABs has been evaluated and discussed with the reference scenario in the previous subchapter. In the following, the operation stability of the half-open ZAB is analyzed by means of the introduced relative humidity scenario, the active operation scenario, the carbon dioxide scenario and the oxygen scenario. The aim of these investigations is to assess which air-compositions should be adjusted to achieve long-term stable ZAB operation.

Evaluation of the Relative Humidity Scenario (b)

Figure 9.3 (a) and 9.3 (b) illustrate the simulation results for the relative humidity scenario for ZAB operation at T=298 K. Different simulations with constant relative humidity in the environment of 0.5, 0.6, 0.65, and 0.7 were conducted. For a better visualization, solely the values at the discharge end of each cycle are plotted. It can be observed that the relative water level will increase if the surrounding relative humidity is above 0.65, and that the relative water level for ZAB operation will decrease if the surrounding relative humidity is below 0.65. This is accompanied by a slight increase and decrease in cell potential for a relative humidity of 0.5 and 0.6, and 0.7, respectively. The cell potential change can be explained with an associated increase and decrease of the overpotentials at both electrodes due to the loss or gain of water, the solvent in the electrolyte. To support this thought, table A.8 in the appendix lists the simulated cell potentials, the overpotentials, and the OCP of the zinc electrode and the air electrode.

Table 9.1 states the values of c_{KOH} that will adjust after 50 and 500 cycles for the simulations with a relative humidity of 0.5 and 0.7, respectively. It can be seen that with RH = 0.5 and increasing cycle number, the KOH-molarity increases. With RH = 0.7 and increasing cycle number, the KOH-molarity decreases. The reason for this behavior is the loss and gain of the solvent water in the electrolyte due to the exchange of water with the surrounding, as given by $J_{H_2O(g)}$. The underlying theory was explained in subchapter 7.1 and shown in figure 7.1 (b): A KOH-molarity different from the initial one will adjust if the electrolyte is exposed to a certain RH.

Since the KOH molarity has a direct influence on the overpotentials at both electrodes, the cell potential simulated increases for lower RH-values, and decreases for greater RH-values than 0.65 due to the change in KOH-molarity. More importantly, it can be observed for the relative humidity scenario that RH = 0.65 evokes a steady state for the cell potential and the relative water content in the air electrode, similar as presented for the reference scenario. This can be explained by the fact that $y_{H_2O(g)}^{sat}$ in the air electrode is equal to the environmental $y_{H_2O(g)}^{env}$ at approximately RH = 0.65 (see also subchapter 7.1 and equation (7.5)). Thus, the net molar flow rate of gaseous water exchange is zero at this relative humidity.

Table 9.1.: Simulated concentrations in the air electrode for the relative humidity scenario at the end of 50 cycles (simulation from figure 9.3), at the end of 500 cycles, and theoretical calculation from equation (7.5) for pure KOH-solution.

	RH=0.5	RH=0.7
c_{KOH}^{air} after 50 cycles	$7.03 \ mol \cdot l^{-1}$	$5.73 \ mol \cdot l^{-1}$
c_{KOH}^{air} after 500 cycles	$7.94 \ mol \cdot l^{-1}$	$5.41 \ mol \cdot l^{-1}$
theoretical/equilibrium c_{KOH}	$7.95 \ mol \cdot l^{-1}$	$5.35 \ mol \cdot l^{-1}$

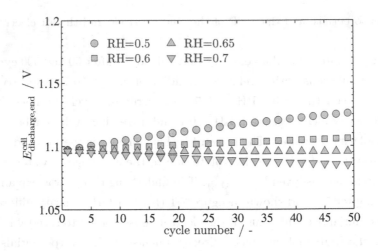

(a)

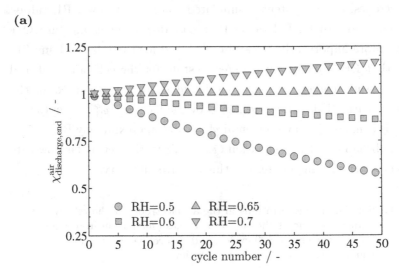

(b)

Figure 9.3.: Simulation results for the relative humidity scenario obtained with different RH-values: (a) cell potential E^{cell} at the end of each discharge cycle; (b) relative water content χ in the air electrode at the end of each discharge cycle.

All in all, it is evident that the RH required to maintain a constant water level in ZABs, is primarily determined by external factors, such as temperature and relative humidity of the surrounding. These external factors can thus influence the cell state, e.g. the water and electrolyte amount, and the KOH-molarity applied inside the battery.

A more abstract interpretation of the simulation results for the relative humidity scenario is presented in figure 9.4. The plot is obtained for running the aforementioned simulations until either $E^{cell} > 2.2$ V, $E^{cell} < 0.9$ V, $\chi^{air} \vee \chi^{zinc} > 1.5$, or $\chi^{air} \vee \chi^{zinc} < 0.5$ is reached, and extracting the number of achieved cycles. The respective result obtained is called operation envelope in the following.

The operation envelope for stable ZAB operation regarding the relative humidity is shown with a gray colored area, and can be interpreted as follows: If the preferred cycle number for ZAB operation is for example 125, it is beneficial to adjust RH-values between 0.55 and 0.75. Any other RH outside the gray colored area would, assumed that this RH remains constant during operation, yield lower cycle numbers than the desired one. Thereby, either cell potential or water level would then become lower than the defined termination values for cell potential and relative water content in the simulation.

Since the previously discussed result is only valid for one distinct KOH-molarity, an extension of the operation envelope is presented in figure 9.5. There, the initial molarity of the alkaline KOH-electrolyte is varied and the maximal cycle number is determined for the operation envelope for each set RH-value. Here, the number of achieved cycles is indicated by the color gradient, depicted in the colorbar on the right-hand-side of the graph. It can be observed that longer operation times, i.e. greater cycle numbers, are obtained with a combination of either elevated RH and low molarity, or low RH and elevated molarity of the KOH-solution. Similar to the operation envelope presented in figure 9.4, a narrow tunnel-like corridor emerges for large cycle numbers (indicated in black) and only certain

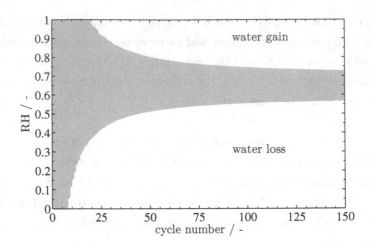

Figure 9.4.: Operation envelope for the relative humidity scenario; constant current cycle simulations for 6 M KOH-electrolyte with termination at either $E^{cell} > 2.2$ V, $E^{cell} < 0.9$ V, $\chi^{air} \vee \chi^{zinc} > 1.5$, or $\chi^{air} \vee \chi^{zinc} < 0.5$.

combinations of RH and initial molarity yield highest cycle numbers. It is apparent that elevated molarities shift the RH needed for large cycle numbers to lower values, which can be explained by the hygroscopic nature of the KOH-solution: High molar KOH-solutions tend to gain water from the surrounding gas phase if it possesses intermediate or high RH-values, as described by equations (7.4) and (7.5).

All in all, the initially chosen liquid electrolyte concentration strongly impacts the water loss or gain for ZABs. The shown results yield for $T^{env} = 298$ K and the commonly applied KOH-molarity of 6 M that lower RH^{env}-values than approximately 0.65 cause water loss, and RH^{env}-values above approximately 0.65 cause water gain for ZAB operation. Associated with the latter is an increase in cell potential, due to a lower water content (concentrated liquid electrolyte), and a decrease in cell potential due to a

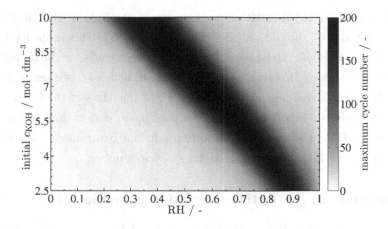

Figure 9.5.: Relative humidity and concentration operation envelope for
the relative humidity scenario at various RH and initial KOH-
molarities; constant current cycle simulations with termination
at either $E^{cell} > 2.2$ V, $E^{cell} < 0.9$ V, or $\chi^{air} \vee \chi^{zinc} > 1.5$, or
$\chi^{air} \vee \chi^{zinc} < 0.5$.

greater water content (diluted liquid electrolyte), respectively. Although
low RH-values cause a short-term cell potential increase, it might not be
favorable for long-term stable ZAB operation to obtain values of χ below
0.5.

All shown cycle numbers are an indicator for the operating time (one cycle
equals 9 hours of battery charge and discharge), which allows to estimate
the duration a ZAB can be exposed to the surrounding RH until it is dried
out or flooded. By implication, it is crucial to either seal the battery very
well or to store it at equilibrium RH according to molarity of the KOH-
electrolyte applied. To prevent water from evaporating during operation,
usually gelling agents are applied in ZABs to bind the solvent of the liquid
electrolyte [113, 114, 115]. However, the electrochemical performance and
the available battery capacity is lower compared to ZABs with liquid alkaline
electrolyte.

It is to be noted that KOH precipitates in the bulk of the electrolyte at a molarity of about 15 at 298 K (see figure 7.5). This molarity is not reached in the bulk electrolyte of zinc and air electrode for the simulation results presented. However, salt precipitation, which subsequently blocks the GDL pores and cuts off oxygen supply, at the edges of the GDL in the air electrode is a reported issue and might appear earlier than in the bulk solution. Thus, the simulation results might not be in line with practical values for the ZAB charge and discharge cycle number.

Evaluation of the Active Operation Scenario (c)

Figure 9.6 illustrates the simulation results for the active operation scenario. Thereby, simulations were conducted for various sets of RH and λ_{O_2}, and then the relative water level χ in the air electrode is evaluated after four constant current density charge and discharge cycles and depicted in the figure. First of all, it can be seen that, similar to the results shown in figure 9.3 (b), the relative water level in the air electrode is increased or diminished according to the set RH. Any RH other than RH^{equi}, will change the water level as indicated with the relative humidity scenario results. Second of all, the change in relative water level in the air electrode due to the RH is increased significantly with increasing λ_{O_2}. The larger λ_{O_2} is, the larger is the flow rate of gas entering the air electrode. Consequently, a significant amount of water vapor can be exchanged with the electrolyte of the ZAB and is either a loss or a gain for the water level (see equation (8.30)). In other words: The closer the RH in the surrounding is set to RH^{equi}, the more independent is the relative water level from the λ_{O_2}-value adjusted.

These findings imply that it is preferable to adjust low λ_{O_2}-values and intermediate RH-values to avoid greater water gain or loss during ZAB operation with 6 M KOH. If active supply of gas is intended for future applications of ZABs, it becomes important to carefully adjust or even control RH and λ_{O_2} according to the actual present electrolyte concentration inside the ZAB.

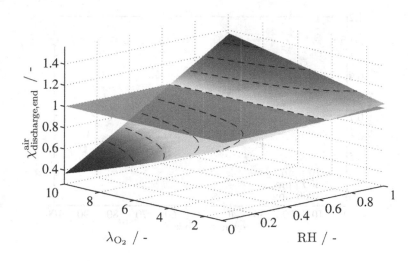

Figure 9.6.: Active operation scenario simulation results: χ in the air electrode for four constant current density charge and discharge cycles. The dash-dotted lines emphasize the curvature of the plot and the gray plane indicates a value of $\chi_{\text{discharge,end}}^{\text{air}} = 1$.

Evaluation of the Carbon Dioxide Scenario (d)

Figure 9.7 (a) and 9.7 (b) illustrate the impact of carbon dioxide on the operation stability of ZABs. In the simulations for these plots, the partial pressure of carbon dioxide in the surrounding air, $p_{CO_2}^{\text{env}}$, is set constant, and is expressed with ppm concentrations at one atmospheric pressure (μl gas per one l of air). The species concentrations in the air electrode obtained for the simulations with 10 ppm and 350 ppm are presented in figure 9.7 (b). It can be observed that the OH^--concentration decreases in the air electrode with increasing ppm of CO_2 and cycle number, which is due to the enhanced replacement by CO_3^{2-}. Likewise, an increase in CO_2-concentration results in a cell potential decrease with operation time. The cell potential decrease can be explained with an associated increase and decrease of the respective overpotentials at both electrodes due to the decrease of OH^- with operation time. To support this thought, table A.9 lists the simulated cell potentials,

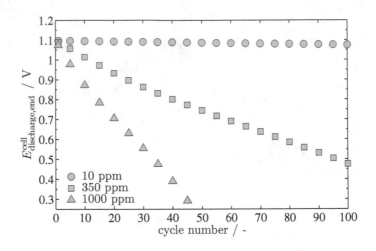

(a)

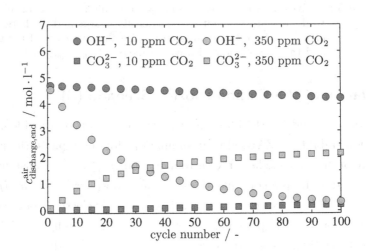

(b)

Figure 9.7.: Simulation results for the carbon dioxide scenario obtained with different $p_{CO_2}^{env}$ in the environmental air: (a) cell potential E^{cell} at the end of each discharge cycle; (b) air electrode molar concentration of OH^- and CO_3^{2-} at the end of each discharge cycle.

the overpotentials, and the OCP of zinc and air electrode in the appendix in subchapter A.3.

The impact of the carbonation on the relative water level in the air electrode is shown in figure 9.8. The simulations indicate an increase of the relative water content in the air electrode with ongoing carbonation since water is produced in reaction (V). The increasing water level is also due to another effect, which can be explained with the help of figure 9.7 (b): The OH^--concentration in the air electrode is decreased due to the replacement by CO_3^{2-}. Consequently, CO_3^{2-} migrate and diffuse towards the zinc electrode, increase the respective concentration there, and thus force the water in the zinc electrode to diffuse towards the air electrode. As as consequence of the increasing relative water content, the OH^--concentration is additionally decreased by the accumulation of water, i.e. a dilution of the liquid electrolyte occurs. However, the observed increase of the relative water content in figure 9.8 is rather small, especially if compared to the changes shown for the relative humidity scenario in figure 9.3 (b).

A more abstract interpretation of the simulation results of the carbon dioxide scenario is presented in figure 9.9 with an operation envelope. The plot is obtained for running simulations until either $E^{cell} > 2.2$ V, $E^{cell} < 0.9$ V, $\chi^{air} \vee \chi^{zinc} > 1.5$, or $\chi^{air} \vee \chi^{zinc} < 0.5$ is reached, and extracting the number of achieved cycles for different CO_2-concentrations. The operation envelope is the gray colored area indicated. It implies that for any cycle number desired, at least a CO_2-concentration has to be adjusted that lies within the gray area.

Compared to the relative humidity scenario, the carbon dioxide influence is stronger on the operation stability of ZABs, so that less cycles can be obtained until battery failure (i.e. E^{cell} or χ are unfavorable) is observed in the shown simulations. Interpolating the given trend for the operation envelope linearly, reveals that a CO_2-concentration of 10 ppm, which can be achieved with common CO_2-filters, would yield approximately 315 cycles, or 2835 hours, of stable operation. This leads to the conclusion that highly purified air supply might be a reasonable choice to ensure long-term stable

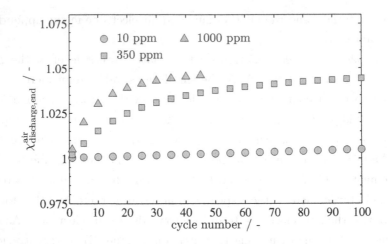

Figure 9.8.: Further simulation results for the carbon dioxide scenario obtained with different $p_{CO_2}^{env}$ in the environmental air, expressed by ppm concentrations for the relative water content χ at in air electrode at the end of each discharge cycle.

ZAB operation. If ZABs would be applied in future automotive applications, they would need to be integrated into a controllable air supply system with filters and blowers, similar as done with fuel cells.

It is to be noted that K_2CO_3 precipitation in the bulk is reached in theory at a molarity of about 6 mol $\cdot$ dm^{-3} (see 7.5). This molarity is not reached for the simulation results depicted. However, salt precipitation of K_2CO_3 might occur locally in the pores of the GDL of the air electrode at any time of operation. The salt might then block the pores for the oxygen transport and cut off the oxygen supply to the CL. Implementing this effect in the carbon dioxide scenario would yield even smaller cycle numbers for the operation envelope.

The model description also allows to change the initial concentration of the alkaline electrolyte and its composition, e.g. the carbonate content. As

With 350 ppm of CO_2, 20 cycles can be reached until the concentration of OH^- has decreased to 2 mol$\cdot$l^{-1} and thus the cell potential is below 0.9 V. This cycle number correlates to an operation time of 180 hours. This value is close to reported experimental findings by [83], where a maximum of 200 hours of stable air electrode operation is reported at 409 ppm of CO_2 until failure is observed. There, the blockage of pores in the air electrode is primarily attributed to the failure. In the simulation results, the cell potential is diminished to 0.9 V after 180 hours mainly by the shortage of OH^- in the electrolyte.

Taking into account that 10 ppm of CO_2 are feasible with carbon dioxide filters (see [83]), the slope of the cell potential for 10 ppm of CO_2 in figure 9.7 (a) is acceptable to maintain a long-term stable ZAB operation. On the contrary, 350 ppm of CO_2, as found in environmental air, and greater ppm-values might not be favorable for a long-term stable operation of ZABs, since the OH^--concentration is decreased drastically; the performance of ZABs is much worse for lower OH^--content in the electrolyte, which was shown experimentally in figure 5.1 (a), and for elevated carbonate amounts in the electrolyte as shown in figure 5.1 (b).

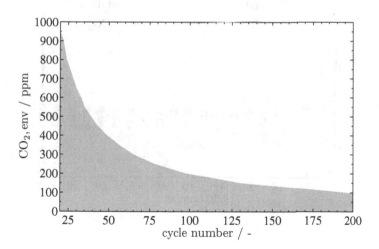

Figure 9.9.: Operation envelope for the carbon dioxide scenario. Simulation results obtained with constant current cycles simulations with termination at either at either $E^{cell} > 2.2$ V, $E^{cell} < 0.9$ V, $\chi^{air} \vee \chi^{zinc} > 1.5$, or $\chi^{air} \vee \chi^{zinc} < 0.5$.

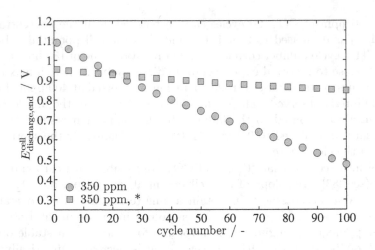

(a)

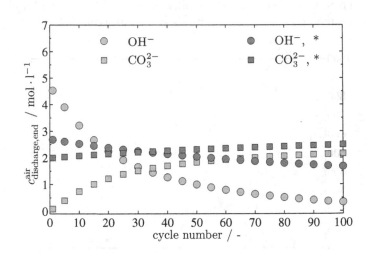

(b)

Figure 9.10.: Simulation results for the carbon dioxide scenario obtained with different initial concentrations for 350 ppm CO_2 at discharge cycle end: Standard case, and carbonate case (*): (a) cell potential E^{cell}; (b) air electrode concentration of OH^- and CO_3^{2-}

described in chapter 7, the reaction rate constant of the carbonate reaction is much lower for lower OH^--concentrations (see figure 7.2). An alternative operating strategy for ZABs can be derived on this basis: Reducing the amount of OH^- in the electrolyte, and adding an appropriate anion, while maintaining the concentration of K^+. To investigate this operating strategy, a simulation for a battery with a mixture of $8.00 \; mol \cdot dm^{-3} \; K^+$, $2.00 \; mol \cdot dm^{-3} \; CO_3^{2-}$, $0.66 \; mol \cdot dm^{-3} \; Zn(OH)_4^{2-}$ and $2.68 \; mol \cdot dm^{-3}$ OH^- ions exposed to 350 ppm CO_2 is compared to the simulation of a battery with standard electrolyte, as described at the beginning of this subchapter, exposed to 350 ppm CO_2. Figure 9.10 depicts the respective simulation results for the cell potential and the ion concentrations in the air electrode.

For cycle numbers below 20, the cell potential obtained is smaller for the simulation with lower initial concentration of OH^-. After 20 cycles, the impact of the carbonation reaction on the cell potential obtained with greater initial concentration becomes severe, and evokes a steep decay of the cell potential. On the contrary, the simulated battery with added carbonates in the electrolyte can maintain an almost constant cell potential level at 0.95 V. In addition, the simulation with lower initial concentration of OH^- yields greater OH^--concentrations at the end of 100 cycles; the low initial concentration slows down the carbonation reaction, since less OH^-, and more of the product CO_3^{2-} is present in the electrolyte. Consequently, the maximal number of (quasi)stable charge and discharge cycles for ZAB operation might be increased significantly. Under this premise, the previously stated model-based results in figure 9.10 (b) and the experimental investigations in this thesis confirm that a novel operating strategy for long-term and stable ZAB operation might be applied: Adding carbonates to the electrolyte solution of ZABs with the intention to slow down the carbonation reaction.

It is to be noted that the overpotentials at zinc and air electrode, and the solubility of oxygen in the electrolyte are not corrected for the presence of CO_3^{2-} ions in the model description of the carbon dioxide scenario. Thus, the cell potential for practical ZAB operation might be different for the simulations shown with changing concentration of OH^- and CO_3^{2-}. For instance, the experimental results in figure 5.1 (b) have revealed that the performance of ZABs operated with KOH-solution with small amounts of carbonates added is acceptable. However, for 50 mol% of added carbonates in the electrolyte, the cell potential during discharge was impacted significantly, as shown in figure 5.4.

Evaluation of the Oxygen Scenario (e)

The following simulation results are conducted with initial conditions that differ from the ones used in the aforementioned scenarios, e.g. a fully charged ZAB with the maximal amount of Zn is initialized (see table A.2). The respective results shown in figure 9.11, can be understood as polarization curves, as explained in subchapter 3.1. Each curve was obtained by constant current discharge simulations with a holding time of 60 seconds. Before each constant current discharge, the simulation initials from table A.2 were restored to simulate each time with a fully charged battery. The discharge steps evaluated start from $0 \, mA \cdot cm^{-2}$ and increase in steps of $0.5 \, mA \cdot cm^{-2}$. Each cell potential at the end of one step is allocated to the respective input discharge current density. The cell potential at the end of the holding time is taken to plot the polarization curves depicted. The results shown in figure 9.11 contain polarization curves for different oxygen partial pressure, p_{O_2}, of 0.0213 MPa in air and 0.1013 MPa with pure oxygen, as well as a polarization curve obtained with the reference scenario model. For the reference scenario, an all time sufficient supply of oxygen is implemented according to its solubility in the KOH-electrolyte. Consequently, the polarization curve obtained for the reference scenario does not possess a distinct steep decline of the cell potential at large current densities because no mass transport limitation of oxygen occurs. The zinc electrode is thereby also not limiting since sufficient amount Zn is implemented.

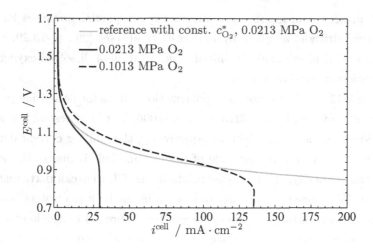

Figure 9.11.: Cell potential E^{cell} for polarization curve simulations for the reference scenario and the oxygen scenario with different oxygen partial pressure p_{O_2} of 0.0213 MPa and 0.1013 MPa, assuming passive oxygen supply.

The results of the oxygen scenario simulations reveal that the limiting current density for ZAB operation increases with p_{O_2}, namely from about 29 mA $\cdot$ cm^{-2} for 0.0213 MPa to about 135 mA $\cdot$ cm^{-2} for 0.1013 MPa partial pressure of oxygen. Experimentally determined polarization curves for full ZABs by Müller et al. [4] possess the same shape and trend for the limiting current density. Additionally, a difference in OCP between the simulations with 0.0213 and 0.1013 MPa p_{O_2} of about 15 mV can be seen in the simulation data, which implies that the OCP is larger for elevated oxygen partial pressures. This observation concurs with the trend given by the Nernst potential for the air electrode at OCP (see equation (A.29)).

During the operation of ZABs in portable appliances, the oxygen partial pressure can change significantly with altitude, and thus the amount of oxygen that can dissolve in the electrolyte. For the following simulation results presented in figure 9.12, various oxygen partial pressures for the respective altitude, as given by Passaniti et al. (see [29], p. 13.29, table 13.7), are used to calculate the initial concentration of dissolved oxygen for polarization curve simulations.

Figure 9.12 (a) shows simulated polarization curves for decreasing partial pressures of oxygen from 0.0213 MPa to 0.00625 MPa to resemble a ZAB operation with increasing altitude. Apparently, the limiting current density decreases with increasing altitude of operation. This is due to the lower concentration of oxygen in the electrolyte at the CL that can participate in reaction (IV), since the solubility of oxygen in the bulk electrolyte is lower with decreasing partial pressure of oxygen (see figure 7.3 in subchapter 7.3). Furthermore, figure 9.12 (b) illustrates on the one hand extracted values for the current densities at 0.9 V from figure 9.12 (a), and on the other hand experimental results for the current densities obtained at 0.9 V for a PR48-type ZAB button cell operated at a pressure equivalent to the respective altitudes (see [29], p. 13.30). It can be observed that a change in altitude from low to high, e.g. from London, UK (0 m altitude) to Mt. McKinley, USA (6082 m altitude) will cause a change of approximately 50% for the ZAB limiting current density. This figure can be understood as operation envelope for the altitude operation of ZABs: The higher the altitude, the lower the current density that can be obtained from the ZAB.

Besides, a deviation of approximately 34% between the current densities from simulation and experiment at low altitudes is observed. One possible explanation for this observation is as follows. The model is considering free diffusion through a highly porous GDL with a rather ideal approach, whereas the experiments are with a PR48-type ZAB button cell that possesses a multitude of PTFE membranes on top of the air electrode's GDL to keep $H_2O(g)$ and CO_2 inside or outside the ZAB, respectively. The diffusion of O_2 into the air electrode is diminished by these PTFE membranes, and thus less oxygen is present directly at the CL, so that presumably less current can be withdrawn from the battery.

In summary, the amount of oxygen in the supplied air for ZABs is a crucial impact factor, especially for portable operation of ZABs where a change in altitude or partial pressure of oxygen can occur. This implies that the use of pure oxygen with a constant partial pressure should be considered for ZABs to achieve larger cell potential and limiting current densities.

All in all, the simulation results presented in the previous subchapters reveal that the introduced model is capable to describe the ideal discharge and charge process of ZABs, since the obtained results are in line with reported findings in literature, and with the experimental findings for ZAB discharge presented in chapter 5. Furthermore, the simulation results have revealed that the surrounding air has indeed an impact on the long-term stable operation of ZABs, and can strongly decrease the cell potential and the available cycle numbers for the charge and discharge of ZABs. Operation envelopes for the stable operation of ZABs for the relative humidity, the carbon dioxide and the oxygen applied have been elucidated. With these envelopes it is possible to assess which air-compositions should be adjusted, and which electrolyte should be chosen to achieve long-term stable ZAB operation.

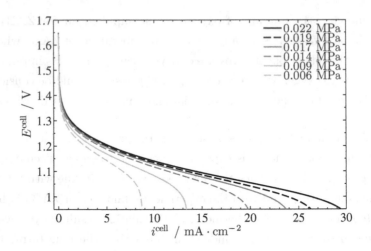

(a)

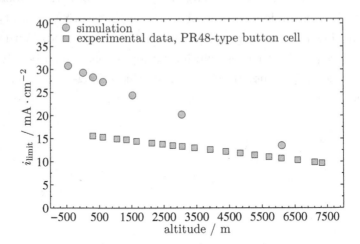

(b)

Figure 9.12.: Further simulation results for the oxygen scenario with different oxygen partial pressures: (a) various polarization curves with decreasing partial pressure of oxygen; (b) current densities at 0.9 V with simulations and experiments (see [29], p. 13.30).

9.3. Validity of the Model-Based Analysis

The investigations presented in the previous subchapters are meant to be an ideal case approach to elucidate the air-composition impacts and the thereof emerging limitations of ZAB operation. In the following, two causes, one being of physical nature and the other being of chemical nature, for possible limits of the validity of the presented models and simulations are discussed.

9.3.1. Unaccounted Processes

Certain processes known to take place, e.g. for real ZAB operation, are not included in the model description presented in this chapter. Two possible processes and their impact on the simulation results presented will be addressed briefly in the following.

Electrode Flooding with Electrolyte

The effect of GDL Flooding with electrolyte and a possible electrolyte depletion at the CL are not accounted for in the presented models introduced in subchapters 8.2 and 8.3. In real ZAB applications, liquid electrolyte in the pores of the GDL would increase the mass transfer resistance for dissolved oxygen to reach the reaction zone. Thus, the overpotential at the air electrode simulated for the relative humidity scenario, which is linked to the oxygen concentration in the air electrode, might be different in practical ZAB operation. This issue was shown experimentally with the discharge of the in-house set-up (see figure 5.11 and figure 5.13), and further investigated with the extended air electrode model in chapter 6.

By implication, values for χ^{air} and χ^{zinc} below 1.0 and above 1.0 might not give practical values for the presented cell potentials and the implications derived from them, if flooding is taken into account. It becomes apparent that on the one hand an increase of the liquid volume in the GDL pores would cause a higher liquid film thickness and consequently a greater diffusion resistance for dissolved oxygen, representing a lower achievable current density or cell potential. On the other hand, a change in electrolyte composition and water level with increasing cycle number, might not diminish the battery performance as shown with the relative humidity and carbon dioxide scenario simulation results. For instance, a higher flooding level in the GDL with more liquid electrolyte volume would imply for the carbon dioxide scenario slightly prolonged operation times because a higher amount of hydroxide ions is present in the electrolyte.

Electro-Osmotic Water Drag

The electro-osmotic drag (EOD) of water can be seen as the simultaneous transport of water molecules alongside with the transport of hydroxide ions due to diffusion and migration. The influence of the EOD of water is discussed widely, for example in alkaline fuel cell research. A value of $k_d = 4$ for the dimensionless water drag coefficient is reported in alkaline solutions and anion exchange membranes [116].

In the case of ZAB operation, the EOD of water seems not to be essential, as it might be compensated by convection and diffusion. This can be observed for instance with the reference scenario during ZAB discharge. There, the hydroxide ions would migrate into the air electrode and would drag the water molecules alongside with them. Since no pressure increase in the zinc electrode is allowed (see equation (8.11)) and the liquid volume in the zinc electrode can not further take up water molecules, the water molecules entering cause a counter flow of liquid electrolyte from the zinc electrode to the air electrode. This will change the concentrations in the respective electrode, which is consequently balanced by the diffusive fluxes described. To demonstrate the impact of the EOD in ZABs, the EOD for water is implemented in the basic model equations, as described in the appendix in equation (A.27).

Figure 9.13 (a) and figure 9.13 (b) show simulation results for constant current density charge and discharge cycles that are obtained with the

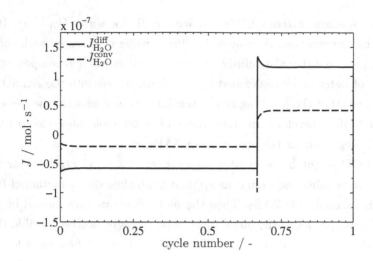

(a)

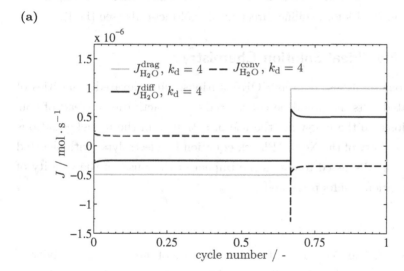

(b)

Figure 9.13.: Simulation results for the molar flow rates of water for: (a) one cycle obtained with the reference scenario without electro-osmotic drag; (b) one cycle obtained with the reference scenario with $k_d = 4$.

reference scenario without EOD and with EOD for water ($k_d = 4$). It is shown that, as expected, the convective flux of water is increased with added EOD impact, and that the diffusive molar flow rate of water is compensating the flow of water molecules caused by the EOD. A look into the simulation data reveals that the resulting cell potential is almost exactly the same as without EOD. Therefore, the movement of water molecules due to EOD might be neglected for the operation of ZABs.

The EOD might be of greater importance if solid-electrolytes or ion conductive membranes, similar as applied in alkaline direct-methanol-fuel-cells, will be applied in ZABs. Then the molar flow rates are limited by the transport properties of the membranes, and it might not be possible that the EOD can be compensated for. Thus, the EOD might be crucial for the water management in electrochemical cells with membranes, as implied by simulation results for alkaline direct-methanol-fuel-cells (see [117]).

9.3.2. Non-Ideal Solution Chemistry

In aqueous solutions, such as KOH-solutions, with elevated molarities of dissociated salts, it should be considered to use activities instead of concentrations. In the following, the use of activities in the model equations and the validity of the Nernst-Planck equation for electrolytes with elevated molarity are discussed to assess the impact of their use on the validity of the simulation results presented.

Activities

In strong alkaline electrolytes, two activities are of interest: The activity of the solute in presence of the dissociated ions with their ionic strength, and the activity of the solvent, here water. The single activity of an ion can not be measured, so that mean ionic activities and the respective coefficients are commonly applied instead (see [118], p. 18).

The use of activities instead of concentrations is neglected in this thesis, because not all species activity coefficients and their influence on each other (e.g. for the quaternary system H_2O-K^+-OH^--$Zn(OH)_4^{2-}$) are available in literature. However, experimental data on the activity of the solute KOH and the solvent water is given (see [119] and [80], respectively).

To link the concentration of species k to its activity a_k, the following expression can be applied (see [118], p. 17):

$$a_k = \gamma_k(c_{KOH}) \cdot \frac{c_k}{c^{ref}} \tag{9.1}$$

where $k = H_2O$ or KOH, and γ is the dimensionless activity coefficient, which depends on the KOH-molarity applied (see [46], p. 52).

On this basis, the mean activity coefficient of KOH and the activity coefficient of water are illustrated in figure 9.14 as a function of KOH-molarity. It is depicted that both activity coefficients are unity for 0 M KOH-solution. This implies, per definition in equation (9.1), that the dimensionless concentration is equal to the activity value at this molarity. With increasing KOH-molarity the activity coefficient for water is decreasing non-linearly, which can be explained with the increased amount of K^+ and OH^- in the solution that decreases the water activity. For 6 M KOH-solution, which is of primary interest for the ZABs in this thesis, the activity coefficient of water is approximately 0.7. Furthermore, it can be observed that the mean activity coefficient of KOH is decreasing quadratically from 0 M to its minimum value of approximately 0.7 at 0.7 M. Subsequently, it is increasing quadratically with a smaller gradient than before with increasing KOH-molarity. For 6 M KOH-solution, the mean activity coefficient of KOH is approximately 2.6. This would imply that the concentration of a 6 M KOH-solution is multiplied by this factor to obtain its actual activity within the solution.

The activity due to the contribution of K^+ and OH^- ions to the electrode reactions, the transport processes, and the empirical equations for the ZAB model approach need then to be scaled respectively. All previously introduced reaction and transport process from subchapters 8.2 and 8.3 would require a correction for the activity. However, the real ion-ion-solvent-

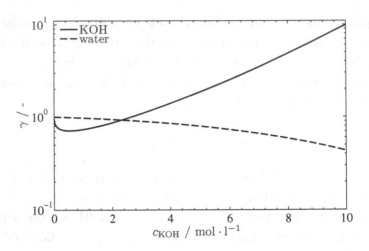

Figure 9.14.: Mean activity coefficient for KOH and activity coefficient of water as a function of KOH-molarity; data as given by [80] and [119], respectively.

interactions that take place in ZABs are even far more difficult to predict and describe with activities than with concentrations for ZABs, since not all activity coefficients are available in literature. Thus, activities are neglected in this thesis. Nonetheless, it is evident that the activity coefficient of water in the electrolyte applied in this thesis is lower under the presence of the high molar KOH-solution, and that the mean activity coefficient of KOH is greater than for an ideal diluted electrolyte system. To estimate the influence of activities on the cell potentials obtained for the reference scenario simulations, the following modifications of the reference scenario are introduced in this subchapter of the thesis: The values for a_{KOH} and a_{H_2O} are implemented in the equation (8.1), (8.3) and the Nernst equations (A.28) and (A.29).

The simulation results for the reference scenario with applied activities, as described above, are illustrated in figure 9.15. The usage of activities generally causes the same charge and discharge cell potential progression with time, but shows in detail an approximately 91 mV larger discharge cell potential than for the reference scenario simulation. The cell potential is approximately 17 mV greater for the reference scenario during charge. Due to the modification of the Nernst potential with activities, the OCP (not shown) of the reference scenario is approximately 11 mV lower for the reference scenario simulations than for the usage of activities.

The general trend for the cell potentials observed in figure 9.15 (a) can be explained with the help of figure 9.15 (b). Therein, the absolute values of the overpotentials at zinc and air electrode for the same simulations with and without activities are shown for the first ZAB charge and discharge cycle. It can be observed that for the charge, η^{air} is greater for the reference scenario than for the evaluation with activities. Thus, a lower loss of potential might emerge due to the charge polarization when considering activities. This is due to the elevated activity of OH$^-$ and the lower activity of H_2O at this KOH-molarity (see figure 9.14). Evidently, they influence the OER (reaction (IV)) so that the charge reaction rate is enhanced and thus the overpotential is decreased. For the discharge, η^{air} for the reference scenario is marginally lower than for the evaluation with activities. The influence on the electrochemical reaction (IV) (now the ORR) is then analogously the other way around. Furthermore, it can be observed that for the charge, η^{zinc} is lower for the reference scenario than for the evaluation with activities. This implies a bigger loss of potential due to charge polarization when considering activities. In this case, it can be explained by the elevated activity of OH$^-$, which is hindering the electrochemical reaction (I). For the battery discharge, the effect is exactly reversed, so that η^{zinc} is bigger for the reference scenario than for the evaluation with activities. This causes a bigger loss of potential due to polarization, and thus a greater cell potential during discharge for the use of activities.

The previously presented considerations are meant to give an outlook on the use of activities instead of concentrations in ZAB models. In particular, the activity impact can be more severe for the other environmental scenarios analyzed and the therein implemented physical processes. The magnitude

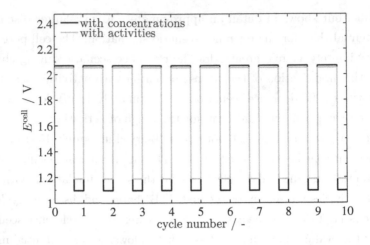

(a)

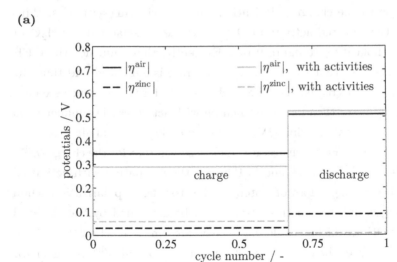

(b)

Figure 9.15.: Simulation results for the reference scenario: (a) cell potential
E^{cell} of reference scenario and reference scenario with mean
KOH-activity and water activity instead of concentrations; (b)
absolute values of the overpotentials at zinc and air electrode
for the same simulations, shown for the first cycle.

of the air-composition impact might change significantly when applying actives. This holds for example for the water vapor equilibrium between the surrounding and the KOH-solution, CO_2-solubility and CO_2-diffusivity, as well as for the O_2-solubility and the O_2-diffusivity. The thoughts and considerations presented on the usage of activities can be the starting point for more detailed and general valid models of ZABs in future work.

Nernst-Planck Equation for Highly Concentrated Solutions

All concentrations gradients applied in equation (8.5) are explicitly valid for infinitely diluted solutions (see [98], p. 271). In electrolyte systems with salt concentrations above 0.1 M, rather the concentrated solution theory and not the diluted solution theory should be applied. The migration, diffusion and convective fluxes in equation (8.5) would then become more general so that multicomponent interaction would be accounted for (see [98], p. 297). This correction is similar to the Maxwell-Stefan equation and is more specifically known as a relation developed by Onsager (see [98], p. 299). In particular, this has to be considered if the solvent, water in the case of liquid alkaline systems, is not the predominant species in terms of molar amount in the electrolyte. However, a multitude of parameters and their dependency from each other need to be identified experimentally for this approach, such as $D_{i,j}$, which is the diffusion coefficient of species i interacting with species j in the solvent.

Psaltis and Farell [120] examine a model-electrolyte, and investigate whether the use of the Nernst-Planck equation or the Maxwell-Stefan approach is to be favored. They report that the transient concentration profiles of the species in the electrolyte were affected by the chosen approach, which in turn influences the time taken to reach steady state for the simulations. However, the steady state concentration profiles of the species in the electrolyte are approximately identical for both approaches. They conclude that if transient profiles are not required, a simplified Maxwell-Stefan model that makes use of binary diffusion coefficients, as found in the

Nernst-Planck equation for diluted solutions applied for this thesis, might be applied. Costly calculations due to the linkage of $D_{i,j}$ and the species concentrations can then be avoided. Finally, they propose to consider the chosen approach carefully only when studying the multicomponent charge transport in nanoporous structures, especially for research problems about the formation and degradation of the solid-electrolyte-interface in lithium-ion batteries where the surface concentrations are of primary interest. This is not the case for the ZAB modeling in this thesis.

Dreyer et al. [121] demonstrate shortcomings in the original description of the Nernst-Planck equation for the thermodynamic coupling of diffusion and mechanics for the species in the electrolyte. As a consequence, physically not meaningful and too large species concentrations close to the electrode surfaces can emerge due to large electrode polarization. Dreyer et al. propose an alternative formulation of the Nernst-Planck equation and successfully apply it on a ternary model-electrolyte. Fuhrmann [122] compares appropriate alternative formulations and their numerical solution to overcome the aforementioned shortcomings. Lumped models, such as the approach in this thesis, are not affected significantly by the inaccuracy of the original Nernst-Planck description [123].

Multicomponent Electrolyte

The KOH-solution applied as electrolyte in ZABs is a multicomponent system of various ions and the solvent water. Thus, it can be assumed that certain material properties will behave differently in practical cases. Most importantly, the solubility of oxygen in the electrolyte might behave differently in real applications. The solubility of oxygen is of special importance, since it has a direct influence on the overpotential at the air electrode and thus the limiting current density for ZABs. Hence, the cell potentials simulated and predicted for the relative humidity scenario and the carbon dioxide scenario might not be in the same region as for practical ZAB applications.

Since the KOH-solution applied as electrolyte in ZABs is a mixture composed of the solvent H_2O, the two ions K^+ and OH^-, and possibly the ion $Zn(OH)_4^{2-}$, it is of ternary nature, respectively quaternary nature. Kriegsmann et al. [124] report the error for a discharge profile for a cylindrical alkaline battery model obtained for a binary and a ternary electrolyte. They show that the model with ternary electrolyte is prone to cause numerical issues, implies greater computational costs, and is less accurate when compared with experimental discharge profiles. Besides, they bring to mind that much more experimental data is available for the binary KOH-electrolyte. They conclude that the model evaluated with binary electrolyte is more efficient, is based on a more understandable model formulation, and yields more opportunities to address certain phenomena in the alkaline cell such as zinc electrode expansion. These are also valid points for the ZAB modeling in this thesis.

In summary, this thesis considers solely the infinitely diluted binary electrolyte. However, non-ideal behavior could be implemented for further detailed model-based investigations of the processes in the zinc electrode and the air electrode.

10. Summary and Overall Conclusions

This thesis reveals fundamental insights into the processes in zinc air batteries. Thereby, experimental and model-based investigations were applied to specifically address the diverse physical, chemical, and electrochemical processes on electrode and on full cell level.

10.1. For Part 1

First, the processes inside zinc air batteries were investigated by means of electrochemical measurement methods and X-ray tomography during operation. For each measurement method, two specific in-house battery set-ups were constructed, and commercial battery types were chosen to investigate the processes of interest. The first set-up was designed for the electrochemical analysis of batteries with various electrode compositions. The second set-up was designed for X-ray tomography with the aim to visualize the active material Zn and the liquid electrolyte inside the battery during operation. To support these investigations, button cells were analyzed with electrochemical and X-ray tomography methods.

Polarization curves, impedance responses, and discharge curves of zinc air batteries with various composition were measured. The analysis shows that the processes at both electrodes are interconnected and that they largely influence each other. In particular, the battery state (electrolyte composition, state-of-discharge, catalyst applied) and the operating strategy (discharge current density) strongly affect the electrochemical state of the electrodes in zinc air batteries.

As the electrolyte molarity can be adjusted easily before use as one parameter in practical ZAB operation, the aforementioned results imply that the KOH-molarity is one key variable to achieve improved battery performance for future ZAB applications. It was shown that the usage of elevated molarities is essential to achieve improved polarization performance. Adding carbonates to the KOH-electrolyte, while keeping elevated molar amounts of potassium ions, may still result in acceptable cell performance.

Selected electrochemical results were correlated to effects on electrode and full cell scale by means of in operando X-ray tomography. Thereby, the underlying reaction and transport processes were identified to affect the discharge of zinc air batteries as follows. The electrochemical and chemical reactions in the zinc electrode cause a volume expansion of the solid materials during discharge. The volume expansion of the zinc electrode in turn affects the state of the air electrode significantly, because liquid electrolyte in the zinc electrode is displaced by the formed solid species, and thus is transported into the air electrode. There, the liquid electrolyte floods the pores of the gas diffusion layer so that the transport resistance for the reactant oxygen to the catalyst layer is more and more increased with operation time. This can lead to an early end-of-life for zinc air batteries, which was first shown experimentally, and then substantiated by simulations with a one-dimensional mathematical model of the air electrode. Thereby, the oxygen distribution and the overpotential were analyzed for increased flooding in the gas diffusion layer. This clarified that indeed the transport resistance for the reactant oxygen to the catalyst layer was increased by the flooding, which consequently increased the overpotential of the air electrode, and thus could cause an unwanted cell potential decay for zinc air batteries. The results obtained have to be considered carefully for the operation of electrically rechargeable zinc air batteries: A full discharge until all zinc particles are utilized and thus the liquid electrolyte is drastically flooding the air electrode, might not be favorable if electrical recharge is intended afterwards.

10.2. For Part 2

The experimental investigations from part 1 of this thesis were then extended by studies on the electrically rechargeable zinc air battery. A mathematical model was implemented to first account for the ideal case operation of zinc air batteries with the basic reaction and transport processes observed experimentally, such as the electrochemical conversion of species, the volume expansion, and the liquid electrolyte transport between zinc electrode and air electrode. These processes resemble the essential parts of the charge and discharge operation of zinc air batteries. The analysis indicates that if no side reactions occur during operation, the zinc air battery could be operated for an infinite number of charge and discharge cycles.

Subsequently, the impact of the surrounding air-composition on the half-open zinc air battery was elucidated with theoretical considerations, and with individual scenarios to account for the impact of relative humidity, active supply of the reactant oxygen, carbon dioxide and oxygen in the surrounding air. The investigations show that it is crucial to consider the air-composition impact on electrically rechargeable zinc air batteries, as it affects the cell potential and the maximal cycle number for stable operation of zinc air batteries. On this basis, operation envelopes for the stable operation of zinc air batteries have been elucidated according to the electrolyte applied and the air-composition exposed to. With these envelopes it is possible to assess, which air-compositions should be adjusted, and which electrolyte composition should be maintained to extend the operation time of zinc air batteries. In detail, the underlying processes of water gain or loss through the open air electrode, and the carbonation reaction were identified as the most severe impacts on zinc air battery operation. All in all, the air-composition impacts on the half-open battery and the implications for the cycle stability of electrically rechargeable zinc air batteries were quantified in this thesis as follows:

- CO_2 was identified to have the most severe impact of the investigated components in the surrounding air. Operating zinc air batteries with pure KOH-electrolyte of elevated molarities causes degradation of the electrolyte due to the carbonation reaction. Even with 10 ppm of CO_2 in the supplied air, which can be achieved with standard CO_2-filters on top of the air electrode, the carbonation reaction takes place. The effect is more pronounced for elevated concentrations of CO_2 and becomes unfavorable for 350 ppm, approximately present in ambient air, since it reduces the cycle number drastically, so that the shortage of OH^- becomes significant for the cell potential obtained.

- Intentionally adding approximately 2 mol $\cdot$ l^{-1} (or 25 mol%) of K_2CO_3 to a high molar KOH-electrolyte, containing 8 mol $\cdot$ l^{-1} K^+, can slow down the carbonation reaction. The respective simulations conducted for 350 ppm of CO_2 revealed a significant improvement of the cycle numbers achievable, when compared to the operation with pure 6 M KOH-electrolyte.

- The impact of carbonates on the charge and discharge behavior simulated, can be linked to experimental results obtained in the first part of this thesis. Both, experiment and simulation, show that it is acceptable to add carbonates to the high molar KOH-electrolyte to achieve similar short-term and discharge performance.

- The relative humidity was identified as second most severe impact factor next to CO_2. Thereby, the initially chosen liquid electrolyte concentration strongly impacts the transport of water vapor through the open air electrode, and thus the water loss or gain for zinc air batteries.

- To achieve a reasonable number of charge and discharge cycles for future ZAB applications in humid or dry locations (e.g. Shanghai or Cairo, respectively), it might be essential to match the KOH-molarity applied and the relative humidity present. Depending on the application and the relative humidity in the surrounding, either the relative humidity or the electrolyte composition should be controlled or adjusted to avoid water loss or gain for the battery, and thus to prevent battery failure or unfavorable cell potentials. Nevertheless, additional components, such as humidifiers and concentration sensors, have to be applied, which might decreases the energy density of the entire battery system.

- To avoid greater water gain or loss during the active supply of air, it is important to carefully adjust or even control the relative humidity and the oxygen access factor λ_{O_2} according to the actual electrolyte concentration present in the zinc air battery. This implies that electrically rechargeable zinc air batteries for stationary, portable, or automotive applications would require additional equipment on system level, such as controller units, blowers, concentration sensors, and humidifiers.

- The use of pure oxygen is favorable to obtain greater cell potentials and limiting current densities compared to operation with air. However, purifying would imply additional weight and costs due to the use of additional equipment on system level, such as filters and blowers. The use of a storage tank for pure oxygen might only be reasonable for stationary applications.

The model assumptions for the basic model and the scenarios applied were critically investigated. The assumptions and limitations of the model approach were commented with regard to the current densities applied, the flooding of the air electrode, the electro-osmotic water drag, and the ideally diluted electrolyte theory applied. The models presented could be extended or corrected for non-ideal solution behavior and activities instead of concentrations in future work.

The experimental and model-based techniques applied in this thesis complement each other well and yield a comprehensive understanding of the diverse reaction and transport processes inside zinc air batteries. This thesis has shown that a combination of experimental and model-based techniques can yield the required insight to improve the cycle numbers of zinc air batteries, which might be relevant for their use as next-generation battery for automotive and stationary applications. The introduced analysis is adaptable and can potentially be applied to understand processes in other battery systems. Currently, half-open metal-air batteries, such as lithium-air batteries, are investigated with growing interest. For instance, the impact of the relative humidity on the safety and the cycling performance of lithium-air batteries is investigated [125]. By implication, also the reaction and transport processes in other next-generation metal-air batteries, and their drawbacks due to the air-composition, could be investigated with a combined experimental and model-based approach as presented in this thesis.

Appendices

A. Modeling

In the following, additional information, equations, and derivations for the model-based analysis in this thesis are given.

A.1. Parameters and Derivations for Air Electrode Model

Applied Parameters

The diffusion coefficients for O_2 and OH^- in the KOH-electrolyte with the unit $m^2 \cdot s^{-1}$ are thereby calculated with an empirical expression fitted from data by [76] and [77], respectively, so that:

$$D_{O_2,KOH} = \left(0.47026 + 1.3619 \cdot \exp \left(\frac{-c_{OH}}{3.15255} \right) \right) \cdot 10^{-9} \qquad (A.1)$$

$$D_{OH,KOH} = \frac{k_B \cdot T}{6 \cdot \pi \cdot \eta_{KOH}^{vis} \cdot R_{ion}} \qquad (A.2)$$

$$\eta_{KOH}^{vis} = 0.799533504 \cdot \exp \left(0.155921614 \cdot c_{OH} \right) \cdot 10^{-3} \qquad (A.3)$$

where c_{OH} is in $mol \cdot dm^{-3}$, η_{KOH}^{vis} is the dynamic viscosity of the KOH-solution in $Pa \cdot s^{-1}$, R_{ion} is the effective hydrodynamic radius of the OH^- ion, or Stokes radius, which is chosen to be 4.642×10^{-11} m [77], and k_B is the Boltzmann constant, which is equal to 1.3806×10^{-23} $m^2 \cdot kg^{-1} \cdot s^{-2} \cdot K^{-1}$. The parameters calculated are also given in the publications [72], [73], and [74].

Table A.1.: Applied parameters for the detailed air electrode model.

parameter	value	unit	source
A	1	m^2	chosen
δ_{CL}	1×10^{-7}	m	chosen
δ_{GDL}	250×10^{-6}	m	chosen
D_{O_2,N_2}	2.30×10^{-5}	$m^2 \cdot s^{-1}$	[15]
$D_{O_2,KOH}$	$f(c_{OH})$	$m^2 \cdot s^{-1}$	[76]
$D_{OH,KOH}$	$f(c_{OH})$	$m^2 \cdot s^{-1}$	[77]
ε_{GDL}	0.75	-	chosen
H	$f(c_{OH})$	$mol \cdot m^{-3} \cdot Pa^{-1}$	[76]
T	298	K	chosen
p_{O_2}	0.21×10^5	Pa	chosen
$\hat{c}_{H_2O}$	51.0×10^3	$mol \cdot m^{-3}$	chosen, const.
$\hat{c}_{O_2}$	8.47	$mol \cdot m^{-3}$	chosen
$\hat{c}_{OH}$	6.0×10^3	$mol \cdot m^{-3}$	chosen
$\hat{\eta}$	0	V	chosen
$V_l(t=0)$	1×10^{-9}	m^3	chosen
$dV_l(t)/dt$	1×10^{-12}	$m^3 \cdot s^{-1}$	chosen
α	0.5	-	[85]
C_{CL}	1.40×10^2	$F \cdot m^{-2}$	[85]
k_{forw}	2.37×10^{-9}	$mol \cdot s^{-1}$	chosen
k_{backw}	2.06×10^{-9}	$mol \cdot s^{-1}$	chosen
z_e	2	-	reaction (IV)

Formulation and Spatial Discretization

The following derivations are based on the publication [74]. A finite volume approach is applied in this thesis on the one-dimensional domain $\Omega(x)$ as follows: The domain is discretized into a finite number of connected control volumes using central differential quotients [126]. Nodes for computational evaluation are placed at the center of each control volume, whereas boundary nodes are placed at the center of the control volume faces [126]. A total number of volume elements, n_{total}, is applied. The number of volume elements n in the gas phase and the liquid phase is equal for both phases, and is thus half as much as the total number of finite volume elements n_{total}. The nodes of the volume elements thereby stay equidistant to each other in the respective sub-domain for gas and liquid. However, the distance of the volume elements to each other can be different in the two sub-domains, and is applied as time-variant for both in this thesis. The application of this spatial discretization on the domain $\Omega(x)$ is illustrated in figure A.1 for various times.

In general, the second derivatives in the space domain $\Omega(x)$ are discretized for the volume element k (see [127, 128]), so that:

$$\int_{x_k}^{x_{k+1}} \nabla \left(-D_{j,h}^{\text{eff}} \cdot \nabla c_j \right) \mathrm{d}x \approx -D_{j,h,k}^{\text{eff}} \cdot \frac{c_{j,k+1} - 2 \cdot c_{j,k} + c_{j,k-1}}{\Delta x_k}, \quad (\text{A.4})$$

$$0 \leq k < n_{\text{total}}$$

For the air electrode considered, the step-size of the discretization grid is linked with equation (6.5), and thus with the moving boundary of gas and liquid. To avoid violation of the mass conservation with the applied spatial discretization with moving grid, a molar compensation flux between two neighboring volume elements is accounted for during the discretization (see [128], p. 373-380). If the interface Γ_{lg} moves, for example with increasing

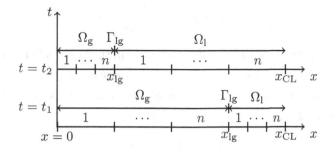

Figure A.1.: Schematic of the spatial discretization in one-dimension with gas filled and liquid filled sub-domains Ω_g and Ω_l. The interface Γ_{lg} is located at x_{lg} and changes its position with time. The number of elements in both sub-domains is n. Reprinted from [74] with permission from Elsevier.

GDE flooding, in negative x-direction to the left-hand-side, a fluid flux must move from one volume element into its right-hand-side neighbor to hold mass conservation in the neighboring volume element. Consequently, the discretized spatial gradient in the liquid filled sub-domain is aggregated, intrinsically due to the nature of the method applied (see [128], chapter 12.4, p. 375-378), by this flux to obtain the final discretization in space of the equations in the one-dimensional domain. The spatial gradient of this flux in the liquid filled sub-domain is applied as suggested by (see [128], p. 376), so that:

$$\int_{x_k}^{x_{k+1}} \nabla \left(\frac{\partial \iota}{\partial t} c_j \right) dx = \left[\frac{\partial \iota}{\partial t} c_j \right]_{x_k}^{x_{k+1}}$$

$$\approx \frac{1}{\Delta x_l(t)} \cdot \left(\frac{d\Delta x_l}{dt} \cdot k \cdot c_{j,k+1} - \frac{d\Delta x_l}{dt} \cdot (k-1) \cdot c_{j,k} \right), \qquad (A.5)$$

$$0 \leq k < n_{total}$$

ι thereby indicates a position on the grid applied. Its derivative by time, $\partial \iota / \partial t$, is a velocity and assesses the grid movement. Both depend on the finite volume element k and its position. Therein, $\Delta x_l(t)$, the size of the liquid filled elements, is described as:

$$\Delta x_l(t) = \frac{V_l(t)}{n \cdot \varepsilon \cdot A} \qquad (A.6)$$

$d\Delta x_l / dt$ is the rate at which one volume element in the liquid filled sub-domain is decreasing or increasing. It is linked with equation (6.5), is expressed as a function of the number of volume elements n, and the porosity of the GDL, ε, so that:

$$\frac{d\Delta x_l}{dt} = \frac{dV_l}{dt} \cdot \frac{1}{n \cdot \varepsilon \cdot A} \qquad (A.7)$$

The final spatial discretization for species j in the liquid filled sub-domain and the volume element k is then:

$$\frac{dc_{j,k}}{dt} = \frac{1}{\Delta x_l(t)} \cdot \left(D_{j,h,k}^{\text{eff}} \cdot \frac{c_{j,k+1} - 2 \cdot c_{j,k} + c_{j,k-1}}{\Delta x_l(t)} \right.$$
$$\left. - ((k-1) \cdot c_{j,k} - k \cdot c_{j,k+1}) \cdot \frac{d\Delta x_l}{dt} \right) + Q_j(t,x) \qquad \text{(A.8)}$$

In the gas filled sub-domain, the concentration is discretized as:

$$\frac{dc_{j,k}}{dt} = \frac{1}{\Delta x_g(t)} \cdot \left(D_{j,h,k}^{\text{eff}} \cdot \frac{c_{j,k+1} - 2 \cdot c_{j,k} + c_{j,k-1}}{\Delta x_g(t)} \right) \qquad \text{(A.9)}$$

where $\Delta x_g(t)$ is expressed as:

$$\Delta x_g(t) = \frac{\delta_{\text{GDL}}}{\varepsilon \cdot n} - \Delta x_l(t) \qquad \text{(A.10)}$$

where δ_{GDL} is the thickness of the entire gas diffusion layer.

Additional Derivations

At the gas-liquid interface Γ_{lg}, Henry's law is considered to describe the solubility of gaseous oxygen in the liquid phase. However, it is a thermodynamic equilibrium expression, which is not suitable for the dynamic pulse-current operation investigated. Therefore, a description to account for the non-equilibrium behavior of the oxygen concentrations in the gas and liquid phase directly at Γ_{lg} is introduced, and used with the following notation: The concentration of oxygen in the last volume element of the gas phase before the interface Γ_{lg}, is denoted as c_{O_2}. The concentration of oxygen in the first volume element in the liquid phase behind the interface Γ_{lg}, is denoted as $c_{O_2}^*$. At the interface Γ_{lg}, $\bar{c}_{O_2}$ is the concentration of oxygen in the gas phase, and $\bar{c}_{O_2}^*$ is the concentration of oxygen in the liquid phase. The relation of the aforementioned concentrations is depicted in figure A.2. Henry's law holds solely at the interface Γ_{lg}. This implies that

$\bar{c}_{O_2}^*$ is expressed as:

$$\bar{c}_{O_2}^* = \bar{c}_{O_2} \cdot H \cdot \mathbf{R} \cdot T \tag{A.11}$$

Henry's law coefficient, H, is expressed in the one-dimensional air electrode model with the following empirical expression, given by Davis et al. [76]:

$$H = \frac{10^{(\log(1.26) - 0.1746 \cdot c_{KOH})}}{p_{atm}} \tag{A.12}$$

Since the conservation of mass has to be accounted for at Γ_{lg}, two flows of oxygen are implemented in the dynamic description of Henry's law. The following molar balance, here expressed with molar fluxes, is applied to describe the latter, so that:

$$N^{out} = N^{in} = -D_{O_2,N_2}^{eff} \cdot \frac{(\bar{c}_{O_2} - c_{O_2})}{\frac{\Delta x_g}{2}} = -D_{O_2,KOH}^{eff} \cdot \frac{(c_{O_2}^* - \bar{c}_{O_2}^*)}{\frac{\Delta x_l}{2}} \tag{A.13}$$

where N is the molar flux of oxygen, and ^{in} and ^{out} indicate the entering and exiting molar flux at the interface. From equations (A.11) and (A.13), $\bar{c}_{O_2}$ can be determined, so that:

$$\bar{c}_{O_2} = \frac{\dfrac{D_{O_2,N_2}^{eff} \cdot c_{O_2}}{\Delta x_g} + \dfrac{D_{O_2,KOH}^{eff} \cdot c_{O_2}^*}{\Delta x_l}}{\dfrac{D_{O_2,N_2}^{eff}}{\Delta x_g} + \dfrac{H \cdot \mathbf{R} \cdot T \cdot D_{O_2,KOH}^{eff}}{\Delta x_l}} \tag{A.14}$$

$c_{O_2}^*$ and c_{O_2} are then calculated according to equation (A.8) and equation (A.9), respectively.

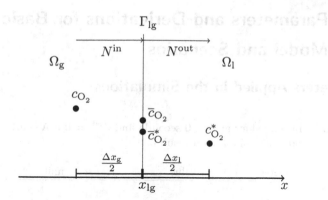

Figure A.2.: Schematic of the concentrations at the interface of gas phase and liquid phase Γ_{lg}. Reprinted from [74] with permission from Elsevier.

A.2. Parameters and Derivations for Basic Model and Scenarios

Parameters Applied in the Simulations

Table A.2.: Initial values at $t = 0$ seconds and $i^{\text{cell}} = 0 \text{ mA} \cdot \text{cm}^{-2}$ for the presented simulation results.

parameter	value	unit
$c_{\text{Zn(OH)}_4^2}^{\text{zinc}}$	0.66	$\text{mol} \cdot \text{dm}^{-3}$
$c_{\text{Zn(OH)}_4^2}^{\text{air}}$	0.66	$\text{mol} \cdot \text{dm}^{-3}$
$c_{\text{K+}}^{\text{zinc}}$	6.00	$\text{mol} \cdot \text{dm}^{-3}$
$c_{\text{K+}}^{\text{air}}$	6.00	$\text{mol} \cdot \text{dm}^{-3}$
$c_{\text{OH}}^{\text{zinc}}$	4.68	$\text{mol} \cdot \text{dm}^{-3}$
$c_{\text{OH}}^{\text{air}}$	4.68	$\text{mol} \cdot \text{dm}^{-3}$
$c_{\text{H}_2\text{O}}^{\text{zinc}}$	50.35	$\text{mol} \cdot \text{dm}^{-3}$
$c_{\text{H}_2\text{O}}^{\text{air}}$	50.35	$\text{mol} \cdot \text{dm}^{-3}$
$c_{\text{O}_2}^{*}$	calculated [63]	$\text{mol} \cdot \text{dm}^{-3}$
$c_{\text{CO}_2}^{*}$	calculated [87]	$\text{mol} \cdot \text{dm}^{-3}$
η^{zinc}	0	V
η^{air}	0	V
$V_{\text{electrolyte}}^{\text{air}}$	6.63×10^{-5}	dm^3
$V_{\text{solid}}^{\text{zinc}}$	$n_{\text{Zn}} \cdot \tilde{V}_{\text{Zn}} + n_{\text{ZnO}} \cdot \tilde{V}_{\text{ZnO}}$	dm^3
$V_{\text{electrolyte}}^{\text{zinc}}$	$1.5 \cdot V_{\text{solid}}^{\text{zinc}}$	dm^3
$\delta_{\text{electrode}}^{\text{zinc}}$	$V_{\text{solid}}^{\text{zinc}}/A_{\text{electrode}}^{\text{zinc}}$	dm
for scenarios (a), (e)		
n_{H}	2.13×10^{-2}	mol
for scenarios (a)-(d)		
n_{Zn}	7.64×10^{-4}	mol
n_{ZnO}	5.50×10^{-3}	mol
for scenario (e)		
n_{Zn}	5.50×10^{-3}	mol
n_{ZnO}	7.64×10^{-4}	mol
for figure 8.5		
n_{Zn}	3.80×10^{-3}	mol
n_{ZnO}	3.10×10^{-3}	mol

Table A.3.: Applied parameters for the basic zinc air battery model. First, geometry parameters are listed, and then material properties are stated.

parameter	value	unit	source
ε^{sep}	0.41	-	chosen
ε^{GDL}	0.5	-	chosen
ε^{air}	0.5	-	chosen
$d_{\text{electrode}}^{\text{zinc}}$	0.13	dm	chosen
$A_{\text{electrode}}^{\text{zinc}}$	1.33×10^{-2}	dm^2	chosen
δ^{sep}	2.50×10^{-4}	dm	chosen
A^{sep}	1.33×10^{-2}	dm^2	chosen
$\delta_{\text{electrode}}^{\text{air}}$	1.00×10^{-2}	dm	chosen
$\delta_{\text{electrode}}^{\text{GDL}}$	2.00×10^{-3}	dm	chosen
$\delta_{\text{film}}^{\text{air}}$	5.00×10^{-7}	dm	[129]
A^{GDL}	7.85×10^{-5}	dm^2	chosen
$A_{\text{electrode}}^{\text{air}}$	1.33×10^{-2}	dm^2	chosen
σ_{Zn}	2.00×10^{6}	$\text{S} \cdot \text{dm}^{-1}$	[9]
σ_{ZnO}	1.00×10^{-9}	$\text{S} \cdot \text{dm}^{-1}$	[9]
$\tilde{V}_{\text{Zn}}$	9.15×10^{-3}	$\text{dm}^3 \cdot \text{mol}^{-1}$	[9]
$\tilde{V}_{\text{ZnO}}$	1.45×10^{-2}	$\text{dm}^3 \cdot \text{mol}^{-1}$	[9]
$\tilde{V}_{\text{Zn(OH)}_4^{2}}$	1.86×10^{-2}	$\text{dm}^3 \cdot \text{mol}^{-1}$	assumed
$\tilde{V}_{\text{CO}_2{}^*}$	3.34×10^{-2}	$\text{dm}^3 \cdot \text{mol}^{-1}$	[130]
$\tilde{V}_{\text{K}_2\text{CO}_3}$	3.25×10^{-2}	$\text{dm}^3 \cdot \text{mol}^{-1}$	[131]
$\tilde{V}_{\text{H}_2\text{O}}$	1.78×10^{-2}	$\text{dm}^3 \cdot \text{mol}^{-1}$	[110]
$\tilde{V}_{\text{OH}}$	7.89×10^{-3}	$\text{dm}^3 \cdot \text{mol}^{-1}$	[132]
$\tilde{V}_{\text{CO}_3^2}$	1.92×10^{-2}	$\text{dm}^3 \cdot \text{mol}^{-1}$	[132]
$C_{\text{DL}}^{\text{zinc}}$	0.020	$\text{F} \cdot \text{dm}^{-2}$	[133]
$C_{\text{DL}}^{\text{air}}$	1.400	$\text{F} \cdot \text{dm}^{-2}$	[134]
$D_{\text{Zn(OH)}_4^2}$	8.55×10^{-9}	$\text{dm}^2 \cdot \text{s}^{-1}$	[13]
D_{OH}	4.94×10^{-7}	$\text{dm}^2 \cdot \text{s}^{-1}$	[135]
$D_{\text{CO}_3^2}$	8.04×10^{-8}	$\text{dm}^2 \cdot \text{s}^{-1}$	[135]
D_{K^+}	1.20×10^{-7}	$\text{dm}^2 \cdot \text{s}^{-1}$	[9]
D_{KOH}	1.20×10^{-7}	$\text{dm}^2 \cdot \text{s}^{-1}$	[136]
$D_{\text{H}_2\text{O}}$	5.26×10^{-7}	$\text{dm}^2 \cdot \text{s}^{-1}$	[98]
$D_{\text{H}_2\text{O(g)}}^{\text{GDL}}$	2.59×10^{-3}	$\text{dm}^2 \cdot \text{s}^{-1}$	[137]

continued

D_{CO_2,N_2}	1.65×10^{-3}	$dm^2 \cdot s^{-1}$	[137]
D_{O_2,N_2}	2.00×10^{-3}	$dm^2 \cdot s^{-1}$	[137]
$D_{CO_2,KOH}$	2.39×10^{-8}	$dm^2 \cdot s^{-1}$	[87]
$D_{O_2,KOH}$	8.00×10^{-8}	$dm^2 \cdot s^{-1}$	[129]
En and Ha	calculated	-	[82]
$c^{sat}_{Zn(OH)_4^2}$	0.66	$mol \cdot dm^{-3}$	[9]
k_I^a	4.94×10^{-11}	$mol \cdot s^{-1}$	*, see text
k_I^c	6.28×10^{-8}	$mol \cdot s^{-1}$	*
k_{II}	0.25	$dm^3 \cdot s^{-1}$	*
k_{IV}^a	2.21×10^{-11}	$mol \cdot s^{-1}$	*
k_{IV}^c	3.49×10^{-12}	$mol \cdot s^{-1}$	*
$k_{CO_2}^{electrolyte}$	1.00×10^{-4}	$dm \cdot s^{-1}$	[138]
$\alpha^{a,zinc}$	0.5	-	[71]
$\alpha^{c,zinc}$	0.5	-	[71]
$\alpha^{a,air}$	0.5	-	[71]
$\alpha^{c,air}$	0.5	-	[71]

Table A.6.: Geometry parameters and initial values for simulations for a ZA13-type button cell.

parameter	value	unit	source
ε^{zinc}	0.5	-	[60]
$d_{electrode}^{zinc}$	0.0580	dm	[60]
$A_{electrode}^{zinc}$	2.64×10^{-3}	dm^2	given by $d_{electrode}^{zinc}$
n_{Zn}	5.59×10^{-4}	mol	[60]
n_{ZnO}	6.14×10^{-4}	mol	assumed

Table A.7.: Matrix of stoichiometric coefficients ν_k and number of exchanged electrons z_e^j.

species	ν_k reaction (I)	(II)	(IV)	(V)	z_e^j
Zn	-1	0	0	0	2
ZnO	0	1	0	0	
$Zn(OH)_4^{2-}$	1	-1	0	0	
OH^-	-4	2	2	-2	
H_2O	0	1	-1	1	
O_2	0	0	$-\frac{1}{2}$	0	2
$CO_2(diss)$	0	0	0	-1	
CO_3^{2-}	0	0	0	1	

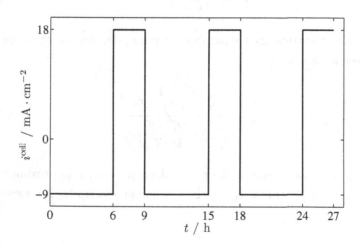

Figure A.3.: Example for the system input for the model-based analysis; three cycles of the constant current density charge and discharge applied.

Diffusion, Migration and Convection in Detail

Equation (8.5) requires the use of molar flow rates of exchange, J, between both ZAB electrodes, namely migration molar flow rates, diffusion molar flow rates and convection molar flow rates. The molar flow rate of exchange for species k in the electrolyte phase is described with the Nernst-Planck equation for diluted solutions, as proposed by [98], so that:

$$J_k^{j,\text{diff}} = D_k^{\text{eff}} \cdot \frac{\left(c_k^{\text{air}} - c_k^{\text{zinc}}\right)}{\delta^{\text{sep}}} \cdot \varepsilon^{\text{sep}} \cdot A^{\text{sep}} \tag{A.15}$$

$$J_k^{j,\text{mig}} = \frac{t_k}{z_k \cdot \mathbf{F}} \cdot i^{\text{cell}} \cdot \varepsilon^{\text{sep}} \cdot A^{\text{sep}} \tag{A.16}$$

$$J_k^{j,\text{conv}} = F^{\text{conv}} \cdot c_k^{\text{zinc}} \tag{A.17}$$

with t_k being the transference number for ion k, determined by properties of all present ions n:

$$t_k = \frac{z_k^2 \cdot \dfrac{D_k^{\text{eff}} \cdot \mathbf{F}}{\mathbf{R} \cdot T} \cdot \check{c}_k}{\sum\limits_n z_n^2 \cdot \dfrac{D_n^{\text{eff}} \cdot \mathbf{F}}{\mathbf{R} \cdot T} \cdot \check{c}_n} \tag{A.18}$$

$\check{c}$, the concentration of species k or n in the separator, is approximated as average concentration of the species in zinc and air electrode. It is expressed as:

$$\check{c} = \frac{c^{\text{zinc}} + c^{\text{air}}}{2} \tag{A.19}$$

In this thesis, the sign for the molar flow rate entering the zinc electrode is denoted as positive so that:

$$J_k^{\text{zinc},h} = -J_k^{\text{air},h} \tag{A.20}$$

Species in Zinc Electrode

The accumulation of zincate ions in the zinc electrode is introduced in equation (A.21) as:

$$
\begin{aligned}
\frac{dc^{zinc}_{Zn(OH)_4^2}}{dt} =\ & \frac{-J^{zinc,diff}_{Zn(OH)_4^2} - J^{zinc,mig}_{Zn(OH)_4^2} - J^{zinc,conv}_{Zn(OH)_4^2}}{V^{zinc}_{electrolyte}} \\
& + \frac{\nu_{Zn(OH)_4^2,I} \cdot r_I + \nu_{Zn(OH)_4^2,II} \cdot r_{II}}{V^{zinc}_{electrolyte}} \\
& - \frac{c^{zinc}_{Zn(OH)_4^2}}{V^{zinc}_{electrolyte}} \cdot \frac{dV^{zinc}_{electrolyte}}{dt}
\end{aligned}
\tag{A.21}
$$

Furthermore, the accumulation of hydroxide ions and water in the zinc electrode is given by the following equations:

$$
\begin{aligned}
\frac{dc^{zinc}_{OH}}{dt} =\ & \frac{-J^{zinc,diff}_{OH} - J^{zinc,mig}_{OH} - J^{zinc,conv}_{OH}}{V^{zinc}_{electrolyte}} \\
& + \frac{\nu_{OH,I} \cdot r_I + \nu_{OH,II} \cdot r_{II}}{V^{zinc}_{electrolyte}} \\
& - \frac{c^{zinc}_{OH}}{V^{zinc}_{electrolyte}} \cdot \frac{dV^{zinc}_{electrolyte}}{dt}
\end{aligned}
\tag{A.22}
$$

$$
\begin{aligned}
\frac{dc^{zinc}_{H_2O}}{dt} =\ & \frac{-J^{zinc,diff}_{H_2O} - J^{zinc,conv}_{H_2O}}{V^{zinc}_{electrolyte}} + \frac{\nu_{H_2O,II} \cdot r_{II}}{V^{zinc}_{electrolyte}} \\
& - \frac{c^{zinc}_{H_2O}}{V^{zinc}_{electrolyte}} \cdot \frac{dV^{zinc}_{electrolyte}}{dt}
\end{aligned}
\tag{A.23}
$$

Species in Air Electrode

The accumulation of hydroxide ions in the air electrode is introduced as:

$$\frac{dc_{OH}^{air}}{dt} = \frac{+J_{OH}^{air,diff} + J_{OH}^{air,mig} + J_{OH}^{air,conv}}{V_{electrolyte}^{air}}$$
$$+ \frac{\nu_{OH,IV} \cdot r_{IV}}{V_{electrolyte}^{air}} \tag{A.24}$$
$$- \frac{c_{OH}^{air}}{V_{electrolyte}^{air}} \cdot \frac{dV_{electrolyte}^{air}}{dt}$$

Furthermore, the change of zincate ions in the air electrode is expressed by:

$$\frac{dc_{Zn(OH)_4^2}^{air}}{dt} = \frac{+J_{Zn(OH)_4^2}^{air,diff} + J_{Zn(OH)_4^2}^{air,mig} + J_{Zn(OH)_4^2}^{air,conv}}{V_{electrolyte}^{zinc}}$$
$$\frac{\nu_{Zn(OH)_4^2,I} \cdot r_I + \nu_{Zn(OH)_4^2,II} \cdot r_{II}}{V_{electrolyte}^{air}} \tag{A.25}$$
$$- \frac{c_{Zn(OH)_4^2}^{air}}{V_{electrolyte}^{air}} \cdot \frac{dV_{electrolyte}^{air}}{dt}$$

Using ohms law as explained in [98], it follows that $\Delta\Phi$ is:

$$\Delta\Phi = \frac{-i^{cell} - \dfrac{\mathbf{F}}{\delta^{sep}} \cdot B_1 + \dfrac{\mathbf{F} \cdot F^{conv} \cdot B_3}{\varepsilon^{sep} \cdot A^{sep}}}{\dfrac{\mathbf{F}^2}{\mathbf{R} \cdot T} \cdot B_2} \tag{A.26}$$

with:

$$B_1 = \sum_k z_k \cdot D_k^{eff} \cdot \left(c_k^{air} - c_k^{zinc} \right)$$

$$B_2 = \sum_k z_k^2 \cdot D_k^{eff} \cdot \tilde{c}_k$$

$$B_3 = \sum_k z_k \cdot c_k^{zinc}$$

Electro-Osmotic Water Drag

To account for the electro-osmotic water drag, Faraday's law is applied. Thus, equation (8.5) of the reference scenario is aggregated with the term:

$$J_{H_2O}^{drag} = k_d \cdot \frac{i^{cell} \cdot \varepsilon^{sep} \cdot A^{sep}}{F} \tag{A.27}$$

where k_d is the dimensionless water drag coefficient.

Application of Nernst Potentials

$$E^{0,zinc} = -1.266 \text{ V} + \frac{R \cdot T}{z_e^{zinc} \cdot F} \cdot \ln\left(\frac{\left(\frac{c_{ZnOH_4^2}^{zinc}}{c^{ref}}\right)}{\left(\frac{c_{OH}^{zinc}}{c^{ref}}\right)^4}\right) \tag{A.28}$$

$$E^{0,air} = 0.401 \text{ V} + \frac{R \cdot T}{z_e^{air} \cdot F} \cdot \ln\left(\frac{\left(\frac{c_{O_2}^*}{c^{ref}}\right)^{\frac{1}{2}} \cdot \left(\frac{c_{H_2O}^{air}}{c^{ref}}\right)}{\left(\frac{c_{OH}^{air}}{c^{ref}}\right)^2}\right) \tag{A.29}$$

$$E^{0,cell} = E^{0,air} - E^{0,zinc} \tag{A.30}$$

Solubility of O_2

The solubility of oxygen in KOH is given by Tromans [63], and is applied in this thesis as:

$$c_{O_2}^{*GDL} = p_{O_2}^{GDL} \cdot \left(\frac{1}{1 + 0.102078 \cdot \left(\frac{\tilde{c}_{KOH}}{c}^{ref} \right)^{1.00044}} \right)^{4.308933} \cdot b_{11} \quad (A.31)$$

with

$$b_{11} = \exp \left(\frac{0.046 \cdot T^2 + 203.35 \cdot T \cdot \log \left(\frac{T}{298} \right) - (299.378 + 0.092 \cdot T) \cdot (T - 298) - 20591}{8.3144 \cdot T} \right)$$

with the unit $mol \cdot l^{-1} \cdot atm^{-1}$. Thus, $c_{O_2}^{*GDL}$, the dissolved amount of oxygen in the liquid electrolyte of the air electrode, depends on the KOH-molarity and the oxygen partial pressure in the GDL, $p_{O_2}^{GDL}$.

Absorption of CO_2

The Hatta modulus Ha and the enhancement factor En are applied for the carbon dioxide scenario as time variant and depend on the concentration of OH^-. They are used as given by [82], so that:

$$Ha = \frac{\sqrt{k_V \cdot c_{OH}^{air} \cdot D_{CO_2,KOH}^{eff}}}{k_{CO_2}^{electrolyte}} \quad (A.32)$$

$$En = \frac{Ha}{\tanh(Ha)} \quad (A.33)$$

Henry's law coefficient is expressed with an empirical equation as given by Pohorecki et al. [87], so that:

$$H_{CO_2} = 10^{\left(-(0.0210+0.0740+0.0660-0.0190)\cdot S_{ionic}+9.1229-5.9044\cdot 10^{-2}\cdot T+7.8857\cdot 10^{-5}\cdot T^2\right)}$$
$$\cdot \frac{1}{p_{atm}} \tag{A.34}$$

where S_{ionic} is the ionic strength of the electrolyte with $k = K^+$, OH^-, CO_3^{2-}, $Zn(OH)_4^{2-}$ and is calculated as proposed by [98], so that:

$$S_{ionic} = \frac{1}{2} \cdot \sum_k z_k^2 \cdot c_k \tag{A.35}$$

k_V is thereby calculated with an empirical expression by [86], so that:

$$k_V = 10^{\left(13.635 - \frac{2895}{T} + 0.08\cdot S_{ionic}\right)} \tag{A.36}$$

Further Empirical Equations

The empirical parameters in equation (7.4) are chosen in this thesis as given by [80], so that:

$$b_1 = -1.508 \times 10^{-2} \qquad\qquad dm^3 \cdot mol^{-1} \tag{A.37}$$

$$b_2 = -1.679 \times 10^{-3} \qquad\qquad \left(dm^3 \cdot mol^{-1}\right)^2 \tag{A.38}$$

$$b_3 = 2.258\,87 \times 10^{-5} \qquad\qquad \left(dm^3 \cdot mol^{-1}\right)^3 \tag{A.39}$$

$$b_4 = -1.206 \times 10^{-3} \qquad\qquad dm^3 \cdot mol^{-1} \tag{A.40}$$

$$b_5 = 5.604 \times 10^{-4} \qquad\qquad \left(dm^3 \cdot mol^{-1}\right)^2 \tag{A.41}$$

$$b_6 = -7.823 \times 10^{-6} \qquad\qquad \left(dm^3 \cdot mol^{-1}\right)^3 \tag{A.42}$$

$$b_7 = 35.45 \tag{A.43}$$

$$b_8 = -3343.93 \qquad\qquad K \tag{A.44}$$

$$b_9 = -10.9 \tag{A.45}$$

$$b_{10} = 4.165 \times 10^{-3} \qquad\qquad K^{-1} \tag{A.46}$$

The saturation partial pressure of gaseous water, $p_{H_2O(g)}$, is calculated with temperature dependent correlations given by Stull [81], so that:

$$p_{H_2O(g)} = 10^5 \cdot 10^{\left(4.6543 - \dfrac{1435.264}{T^{env} - 64.848}\right)} \tag{A.47}$$

with $p_{H_2O(g)}$ in Pa.

The applied molal concentrations, which are used in equation (7.4), are converted to molar concentrations by an expression given by Newman (see [98], p. 36, equation 2.17), as:

$$\tilde{c}_{KOH} = \frac{c_{KOH}}{\rho_{KOH} - \sum\limits_{j \neq 0} c_j \cdot M_j} \tag{A.48}$$

where ρ_{KOH} is the density of the KOH-solution, M_j is the molar mass of species j other than the solvent, and the summation does not include the solvent water, which is denoted with 0. ρ_{KOH} is calculated with an expression by Laliberte and Cooper [139], so that:

$$\rho_{KOH} = \frac{1}{\dfrac{w_{H_2O}}{\rho_{H_2O}} + w_{KOH} \cdot B} \tag{A.49}$$

$$B = \frac{w_{KOH} + 0.14542 + 0.002040 \cdot T}{194.85 \cdot w_{KOH} + 407.31} \cdot \exp\left(10^{-6} \cdot (T + 1180.9)^2\right)$$

where T is in degrees, w indicates the respective mass fraction (with $w_{H_2O} = 1 - w_{KOH}$), and ρ_{H_2O} is calculated with an expression by Kell [140] as:

$$\rho_{H_2O} = \frac{(((((-2.8054253 \cdot 10^{-10} \cdot T + 1.0556302 \cdot 10^{-7}) \cdot T - 4.6170461 \cdot 10^{-5}) \cdot T - 0.0079870401) \cdot T + 16.945176) \cdot T + 999.83952)}{1 + 0.01687985 \cdot T}$$

$$\tag{A.50}$$

Table A.8.: Simulated cell potentials, overpotentials and OCP of the electrodes for the relative humidity scenario (simulation in figure 9.3) at the end of 50 cycles.

	RH=0.5	RH=0.65	RH=0.7
E^{cell} [V]	1.126	1.095	1.085
η^{air} [V]	-0.520	-0.511	-0.508
η^{zinc} [V]	0.052	0.089	0.101
$E^{\text{air},0}$ [V]	0.337	0.345	0.347
$E^{\text{zinc},0}$ [V]	-1.360	-1.350	-1.347

Table A.9.: Simulated cell potentials, overpotentials and OCP of the electrodes for the carbon dioxide scenario at the end of 100 cycles. Potential losses due the ionic resistance of the separator are not listed. They are not constant and change with time with the electrolyte molarity, and thus ionic conductivity. They increase from 0.2 mV to approximately 7 mV at 100 cycles for the simulation with 350 ppm CO_2.

	10 ppm	350 ppm
E^{cell} [V]	1.074	0.476
η^{air} [V]	-0.510	-0.506
η^{zinc} [V]	0.108	0.648
$E^{\text{air},0}$ [V]	0.347	0.413
$E^{\text{zinc},0}$ [V]	-1.345	-1.213

A.3. Additional Simulation Results

Table A.8 lists the cell potentials, the overpotentials, and the OCP of the zinc electrode and the air electrode for the simulations with the relative humidity scenario for $RH = 0.5$, $RH = 0.65$ and $RH = 0.7$ after 50 cycles. Potential losses due the ionic resistance of the separator are not listed. They are not constant and change with time with the electrolyte molarity, and thus ionic conductivity, but are always much smaller (values of approximately 0.2 mV) than the overpotentials at the current densities investigated. It is apparent that the OCP of each electrode changes only marginal with increasing relative humidity. This leaves the change in overpotential at both electrodes as reason for the deviation in cell potential for the simulation with small and large relative humidity values. However, the overpotential at the air electrode only increases marginal with decreasing RH (due to the decreased oxygen solubility and due to the increasing KOH-molarity with operation time). This leaves the increase in zinc electrode overpotential and OCP of the zinc electrode (both due to the increasing KOH-molarity with operation time) as reason for the increase in cell potential observed for low values of RH.

Table A.9 lists the same trends for the carbon dioxide scenario. Thereby, the resulting OH^--concentration for the simulation with 350 ppm causes on the one hand a diminished overpotential and a greater OCP of the air electrode (both due to an improved oxygen solubility). On the other hand, it causes a more severe increase in overpotential and a decrease in OCP at the zinc electrode, which is the reason for the comparably small cell potential obtained after 100 cycles with 350 ppm of CO_2.

B. Experimental

In following, more detailed information for the experimental work in this thesis are given.

B.1. Designing the In-House Set-Up for X-ray Tomography

As elucidated in subchapter 3.2, a multitude of factors have to be considered for designing a battery or fuel cell set-up for X-ray tomography. The following summary states the aspects that were considered carefully for the design of the in-house set-up for the tomography in this thesis:

- PTFE is chosen as housing for the battery set-up. This material combines chemical resistivity against the electrolyte, and a very low attenuation. It is primarily composed of carbon and hydrogen bonds, and minor additives with small amount. Even larger thicknesses of several millimeters are possible compared to a metal housing.

- Graphite/PPS is chosen as current-collector material. It is a carbon-containing and sulfur-containing structure, possesses a very low attenuation coefficient, is chemically and electrochemically stable, and electrically conductive.

- The GDL is also made of carbon, so that the same consideration as mentioned above holds.

- The CL at the air electrode is comparably thin (approximately 20 μm), and contains a perovskite catalyst (primarily composed of La) with large attenuation coefficient. However, the CL is thin enough so that other parts of the air electrode are not significantly affected.

- The pores of the GDL are filled with aqueous KOH-solution. The mixture of water and potassium hydroxide is distinguishable from air in the images obtained due to the larger attenuation coefficient.

- The active material of interest in the zinc electrode, Zn, possesses a two order of magnitudes greater attenuation coefficient at the X-ray energies applied than the other components. This gives a good contrast to the other materials applied, e.g. to the housing and to the electrolyte.

Matlab Source Code for X-ray Attenuation Calculation

```
close all; clear all; clc;

%molar mass of elements in [g/mol]
def.M_zn=65.40; def.M_zno=81.39; def.M_zincate=65.40+4*(16+1);
    def.M_oh=16+1; def.M_h2o=18; def.M_o2=32; def.M_koh=56; def
    .M_k=32; def.M_co32=60; def.M_o2=16;
%densities
def.d_zn=7140; %[g/dm^3]
def.d_zno=5606; %[g/dm^3]


%calculation for attenuation at 50 keV
%source: http://physics.nist.gov/PhysRefData/XrayMassCoef/tab3.
    html


%% PTFE
w_c_PTFE=0.24; %mass fraction of carbon in PTFE [-], since
    formula is (C2F4)n and M_C=12 g/mol and M_F=19 g/mol
w_f_PTFE=0.76; %mass fraction of fluorine in PTFE [-]
mu_c=0.1871/100; %mass attenuation coefficient of single
    component carbon [dm^2/g]
mu_f=0.214/100; %mass attenuation coefficient of single
    component fluorine [dm^2/g]
mu_PTFE=w_c_PTFE.*mu_c + w_f_PTFE.*mu_f;
%mass attenuation coefficient of composite PTFE [dm^2/g]
```

```
%value is in line with given table by NIST (table 4)
density_PTFE=2.20*1000; %density of PTFE [g/dm^3]

%% PPS (graphite material)
% polyphenylene sulfide graphite from Eisenhuth
w_c_pps=0.667; %mass fraction of carbon in PPS [-], since
    formula is (SC6H4)n and M_C=12 g/mol and M_F=19 g/mol
w_h_pps=0.037; %mass fraction of fluorine in PPS [-]
w_s_pps=0.296; %mass fraction of sulfur in PPS [-]
mu_c=0.1871/100; %mass attenuation coefficent of single
    component carbon [dm^2/g]
mu_s=0.5849/100; %mass attenuation coefficent of single
    component sulfur [dm^2/g]
mu_h=0.3355/100; %mass attenuation coefficent of hydrogen [dm
    ^2/g]
mu_pps=w_c_pps.*mu_c + w_s_pps.*mu_s + w_h_pps.*mu_h; %mass
    attenuation coefficent of composite PPS [dm^2/g]
density_pps=1.40*1000; %density of PPS [g/dm^3] source:
    manufacturer

%% Al
mu_al=0.3681/100; %mass attenuation coefficent of single
    component Al [dm^2/g]
density_al=2.7*1000; %density of Al [g/dm^3]

%% C
density_c=2.1*1000; %density of amoprhous C [g/dm^3]

%% copper
mu_cu=2.613/100; %mass attenuation coefficent of single
    component Cu [dm^2/g]
density_cu=2.613*1000; %density of Cu [g/dm^3]

%% Zn
mu_zn=2.892/100; %mass attenuation coefficent of single
    component Al [dm^2/g]

%% ZnO
mu_o=0.2132/100;
mu_ZnO=0.5.*(mu_zn) + 0.5.*(mu_o); %ZnO
```

%% KOH
mu_k=0.8679/100;
mu_o=0.2132/100;
w_K_KOH=def.M_k./(def.M_k+1+def.M_o2/2);
w_O_KOH=def.M_o2/2./(def.M_k+1+def.M_o2/2);
w_H_KOH=1./(def.M_k+1+def.M_o2/2);
mu_KOH=0.25.*(w_H_KOH.*mu_h+w_K_KOH.*mu_k+w_O_KOH.*mu_o) +
    0.75.*(2*w_H_KOH.*mu_h+w_O_KOH.*mu_o); %25% KOH or
    approximated 6 M KOH / water solution
density_KOH=1.25*1000; %density of KOH [g/dm^3] source:
    approximated

%% zinc electrode
w_K_KOH=def.M_k./(def.M_k+1+def.M_o2/2);
w_O_KOH=def.M_o2/2./(def.M_k+1+def.M_o2/2);
w_H_KOH=1./(def.M_k+1+def.M_o2/2);

mu_KOH_el=0.25.*(w_H_KOH.*mu_h+w_K_KOH.*mu_k+w_O_KOH.*mu_o)+
    0.75.*(2*w_H_KOH.*mu_h+w_O_KOH.*mu_o); %25% KOH or approx.
    6 M KOH / water solution

mu_zinc_electrode=0.5.*mu_zn + 0.5.*mu_KOH_el;
%mass attenuation coefficent of composite zinc electrode [dm^2/
    g]

density_zinc_electrode=(1.25*1000+def.d_zn)./2;
%mean density of zinc electrode [g/cm^3] source: approximated

%calculation of attenuation coefficents
mu=[mu_PTFE.*density_PTFE mu_pps.*density_pps mu_al.*density_al
    mu_cu.* density_cu mu_zn.*def.d_zn   mu_zinc_electrode.*
    density_zinc_electrode mu_KOH.*density_KOH mu_c.*density_c
    mu_ZnO.*def.d_zno];
%linear attenuation coefficient of several materials [1/dm]

mu_label={'PTFE','PPS','Al','Cu','Zn','Zn/KOH','KOH','C','ZnO'
    };
delta_material=log(2)./mu*100; %thickness of sample to weaken
    incident beam by half [mm], source: Waseda 2011, equation
    (4), page 16

```
[delta_material_sorted ,index]=sort(delta_material);

figure(1)
h=bar(delta_material_sorted ,0.25);
set(gca, 'YScale', 'log')
set(gca, 'YTick', [10^-4 10^-3 10^-2 10^-1 10.^0 10.^1 10.^2
     10.^3 10.^4 10.^5 10.^6 10.^7 10.^8 10.^9 10.^10])
set(get(h(1),'BaseLine'),'LineWidth',2,'LineStyle',':')
ylabel('thickness/mm')
%sample thickness to reduce X-ray intensity by half
set(gca, 'XTick', 1:length(mu_label), 'XTickLabel',mu_label(
     index));
barmap=[0.7 0.7 0.7];
colormap(barmap);


%calculation of attenuation with energy  NIST database for mass
     attenuation coefficents [cm^2/g] and photon energy [MeV]
%source:http://www.nist.gov/pml/data/xraycoef/index.cfm

photon_energy_Al=[0.00100000000000000;  0.00150000000000000 ;
     0.00155960000000000;  0.00155960000000000;
     0.00200000000000000;  0.00300000000000000;
     0.00400000000000000;  0.00500000000000000;
     0.00600000000000000;  0.00800000000000000;
     0.0100000000000000;  0.0150000000000000;
     0.0200000000000000;  0.0300000000000000;
     0.0400000000000000; 0.0500000000000000;
     0.0600000000000000;  0.0800000000000000; 0.100000000000000;
     0.150000000000000; 0.200000000000000; 0.300000000000000;
     0.400000000000000; 0.500000000000000; 0.600000000000000;
     0.800000000000000; 1; 1.25000000000000; 1.50000000000000;
     2; 3; 4; 5; 6; 8; 10; 15; 20];

photon_energy_C =[0.00100000000000000; 0.00150000000000000;
     0.00200000000000000; 0.00300000000000000;
     0.00400000000000000; 0.00500000000000000;
     0.00600000000000000; 0.00800000000000000;
     0.0100000000000000; 0.0150000000000000; 0.0200000000000000;
     0.0300000000000000; 0.0400000000000000;
```

0.050000000000000; 0.060000000000000; 0.080000000000000;
0.100000000000000; 0.150000000000000; 0.200000000000000;
0.300000000000000; 0.400000000000000; 0.500000000000000;
0.600000000000000; 0.800000000000000; 1; 1.250000000000000;
1.500000000000000; 2; 3; 4; 5; 6; 8; 10; 15; 20];

photon_energy_Zn = [0.001000000000000000; 0.001009800000000000;
    0.001019700000000000; 0.001019700000000000;
    0.001031190000000000; 0.001042800000000000;
    0.001042800000000000; 0.001115650000000000;
    0.001193600000000000; 0.001193600000000000;
    0.001500000000000000; 0.002000000000000000;
    0.003000000000000000; 0.004000000000000000;
    0.005000000000000000; 0.006000000000000000;
    0.008000000000000000; 0.009658600000000000;
    0.009658600000000000; 0.010000000000000000;
    0.015000000000000000; 0.020000000000000000; 0.030000000000000000;
    0.040000000000000000; 0.050000000000000000;
    0.060000000000000000; 0.080000000000000000; 0.100000000000000;
    0.150000000000000; 0.200000000000000; 0.300000000000000;
    0.400000000000000; 0.500000000000000; 0.600000000000000;
    0.800000000000000; 1; 1.250000000000000; 1.500000000000000;
    2; 3; 4; 5; 6; 8; 10; 15; 20];   *%[MeV]*

photon_energy_PTFE = [0.001000000000000000; 0.001500000000000000;
    0.002000000000000000; 0.003000000000000000;
    0.004000000000000000; 0.005000000000000000;
    0.006000000000000000; 0.008000000000000000;
    0.010000000000000000; 0.015000000000000000; 0.020000000000000000;
    0.030000000000000000; 0.040000000000000000;
    0.050000000000000000; 0.060000000000000000; 0.080000000000000000;
    0.100000000000000; 0.150000000000000; 0.200000000000000;
    0.300000000000000; 0.400000000000000; 0.500000000000000;
    0.600000000000000; 0.800000000000000; 1; 1.250000000000000;
    1.500000000000000; 2; 3; 4; 5; 6; 8; 10; 15; 20];

%Al
%http://physics.nist.gov/PhysRefData/XrayMassCoef/ElemTab/z13.
    html

data_Al =[1185; 402.200000000000; 362.100000000000; 3957; 2263;
    788; 360.500000000000; 193.400000000000; 115.300000000000;
    50.3300000000000; 26.2300000000000; 7.95500000000000;
    3.44100000000000; 1.12800000000000; 0.568500000000000;
    0.368100000000000; 0.277800000000000; 0.201800000000000;
    0.170400000000000; 0.137800000000000; 0.122300000000000;
    0.104200000000000; 0.0927600000000000; 0.0844500000000000;
    0.0780200000000000; 0.0684100000000000; 0.0614600000000000;
    0.0549600000000000; 0.0500600000000000;
    0.0432400000000000; 0.0354100000000000; 0.0310600000000000;
    0.0283600000000000; 0.0265500000000000;
    0.0243700000000000; 0.0231800000000000; 0.0219500000000000;
    0.0216800000000000];

%C
%http ://physics . nist . gov/PhysRefData/XrayMassCoef/ElemTab/z06 .
    html
data_C= [2211; 700.200000000000; 302.600000000000;
    90.3300000000000; 37.7800000000000; 19.1200000000000;
    10.9500000000000; 4.57600000000000; 2.37300000000000;
    0.807100000000000; 0.442000000000000; 0.256200000000000;
    0.207600000000000; 0.187100000000000; 0.175300000000000;
    0.161000000000000; 0.151400000000000; 0.134700000000000;
    0.122900000000000; 0.106600000000000; 0.0954600000000000;
    0.0871500000000000; 0.0805800000000000; 0.0707600000000000;
    0.0636100000000000; 0.0569000000000000;
    0.0517900000000000; 0.0444200000000000; 0.0356200000000000;
    0.0304700000000000; 0.0270800000000000;
    0.0246900000000000; 0.0215400000000000; 0.0195900000000000;
    0.0169800000000000; 0.0157500000000000];

%Zn
%http ://physics . nist . gov/PhysRefData/XrayMassCoef/ElemTab/z30 .
    html
data_Zn =[1553; 1518; 1484; 3804; 5097; 6518; 8274; 8452; 7371;
    8396; 4825; 2375; 831.100000000000; 386.500000000000;
    211.800000000000; 129; 58.7500000000000; 35.0500000000000;
    253.600000000000; 233.100000000000; 81.1700000000000;
    37.1900000000000; 12.0700000000000; 5.38400000000000;
    2.89200000000000; 1.76000000000000; 0.836400000000000;

```
      0.497300000000000;  0.234100000000000;  0.161700000000000;
   0.114100000000000;  0.095390000000000;  0.0845000000000000;
   0.076950000000000;  0.066560000000000;  0.0594100000000000;
      0.052960000000000;  0.048340000000000;
   0.042350000000000;  0.036340000000000;  0.0336000000000000;
      0.032250000000000;  0.031600000000000;
   0.031380000000000;  0.031750000000000;  0.0333500000000000;
      0.035090000000000];
```

%PTFE
%http://physics.nist.gov/PhysRefData/XrayMassCoef/ComTab/teflon
   .html
```
data_PTFE=[4823;  1672;  760.100000000000;  241.100000000000;
   104.500000000000;  54.0900000000000;  31.4200000000000;
   13.2700000000000;  6.80500000000000;  2.08800000000000;
   0.966700000000000;  0.402500000000000;  0.264700000000000;
   0.213200000000000;  0.188000000000000;  0.163200000000000;
   0.150000000000000;  0.131000000000000;  0.118900000000000;
   0.102700000000000;  0.091870000000000;  0.0838000000000000;
   0.077470000000000;  0.068030000000000;  0.0611500000000000;
      0.054690000000000;  0.049790000000000;
   0.042800000000000;  0.034560000000000;  0.0298100000000000;
      0.026740000000000;  0.024600000000000;
   0.021850000000000;  0.020200000000000;  0.0181100000000000;
      0.017220000000000];
```

```
figure(2)
rectangle('Position',[25,1.2e-4,75,0.95e2],'FaceColor',[.8 .8
   .8])
hold on
set(gca,'YScale','log')
set(gca,'XScale','log')
loglog(photon_energy_C*1000,data_C/100,'k')
loglog(photon_energy_Zn*1000,data_Zn/100,'k--')
loglog(photon_energy_PTFE*1000,data_PTFE/100,'b')
loglog(photon_energy_Al*1000,data_Al/100,'b--')
xlabel('X-ray-energy/keV')
ylabel('attenuation-coefficient/g/dm^{2}')
legend('C','Zn','PTFE','Al')
```

B.2. Preparation of the Electrolyte Solutions

Two different KOH-solutions were initially prepared using KOH-flakes (Sigma-Aldrich, 90% purity, reagent grade) and pure water with a resistivity of 18.2 $M\Omega \cdot$ cm (Millipore). Approximately 20 wt% excess of flakes was added to ensure defined concentrations of OH^-. The molarity of these two solutions was measured by performing titration and the values found out were 6.07 M and 10.33 M, respectively [57]. All samples prepared, were more than 100 ml in quantity to reduce the error in preparation and to reduce the impact of ambient air on the measurement.

The preparation of the solutions with various amounts of carbonate ions was performed by using K_2CO_3 (Sigma Aldrich) as follows. First, a solution of KOH with defined molarity was prepared and titrated to determine the actual concentration of OH^-. After titration, a molar amount of K_2CO_3 was added to achieve the desired concentration of K^+. The molar amount added is expressed as molar fraction of potassium carbonate, $x_{K_2CO_3}$, and is given by the following relation:

$$x_{K_2CO_3} = \frac{n_{K_2CO_3}}{n_{K_2CO_3} + n_{KOH}} \tag{B.1}$$

$x_{K_2CO_3}$ multiplied by 100 and therefore expressed as molar percentage in this thesis.

B.3. Titration of the Solutions

The molar percentage of carbonates was checked with titration measurements as described in the following. The solutions were titrated to get the actual concentrations and these were verified to lie within one percent of the intended value. The indicator used for the titration was methyl orange (Sigma Aldrich), and a 25 ml standard burette with an accuracy of ±0.5 ml was used to perform the titration [57].

Titrations for two different solutions were conducted: KOH-solution and a KOH-K_2CO_3-solution. For the titration of KOH-solution, a 0.5 M basic standard of potassium hydrogen pthalate (KHP) was used. For the titration of the mixture of KOH and K_2CO_3, a standard procedure was followed [141]: To determine the carbonate content in the solution, $BaCl_2$-solution was used to precipitate the carbonate ions and then the solution was titrated, which gave the K^+ concentration , and excluded the carbonate ions. The carbonate content was determined by subtracting this value from the one obtained without the addition of $BaCl_2$ [57].

B.4. Additional Calculations

The percentage of ionic contribution of the separator, if soaked with 6 M KOH-electrolyte and possessing a thickness of 25 μm (for Celgard 3401), to the overall battery resistance from the EIS measurement for 6.07 M KOH-electrolyte is approximated as:

$$\frac{R^*_{sep}}{R^*_{measured}} = \frac{\frac{\delta_{sep}}{\kappa_{6 \text{ M KOH}}}}{R^*_{measured}} = \frac{\frac{0.0025 \text{ cm}}{616 \text{ mS} \cdot \text{cm}^{-1}}}{0.3723 \ \Omega \cdot \text{cm}^2} = 0.01 \tag{B.2}$$

The percentage of ionic contribution of the porous zinc electrode to the internal battery resistance is approximated as:

$$\frac{R^*_{Zn}}{R^*_{measured}} = \frac{\frac{\delta_{Zn}}{\kappa_{6 \text{ M KOH}}}}{R^*_{measured}} = \frac{\frac{0.17 \text{ cm}}{616 \text{ mS} \cdot \text{cm}^{-1}\text{cm}}}{0.3723 \ \Omega \cdot \text{cm}^2} = 0.74 \tag{B.3}$$

Thereby, the resistance for the ion transport is calculated based on the entire thickness of the zinc electrode.

B.5. Additional Measurements with Commercial Button Cells

Measurement Details

Three button cells from GP of the same lot number were operated in a constant temperature chamber (SU-641, ESPEC, Japan) to obtain the impact of temperature on the polarization behavior of ZABs. The temperature in the chamber was thereby set consecutively from 303 K in steps of 5 K to 353 K, and then consecutively from 293 K in steps of 5 K down to 263 K. The batteries were allowed to settle for every temperature adjusted for at least 10 minutes. For each temperature, the cell potential was recorded for 5 seconds consecutively at a set current density of 0 mA $\cdot$ cm^{-2} (OCP condition), 0.98 mA $\cdot$ cm^{-2}, 4.90 mA $\cdot$ cm^{-2}, 9.80 mA $\cdot$ cm^{-2} and 19.59 mA $\cdot$ cm^{-2} with an automated battery test system (Model 4200 desktop, Maccor, USA).

For one button cell from GP with a SOD close to 100%, a series of three EIS measurements was conducted to investigate the impact of temperature on the impedance response of ZABs. The button cell was subjected to an EIS measurement at 2.93 mA $\cdot$ cm^{-2} constant current density and 0.29 mA $\cdot$ cm^{-2} amplitude of alternating current. The button cell was operated in a climate chamber (Memmert HPP 108) at RH = 0.65, and various temperatures of 298 K, 308 K and 318 K were adjusted. Prior each EIS measurement a holding time of 10 minutes at the set temperature was maintained. The duration of one EIS measurement was 50 minutes.

Temperature Impact on Polarization

Commercial ZAB button cells were operated in a temperature chamber at various current densities. The measurement results are shown in figure B.1. The OCP increases slightly with increasing temperature. For increasing current densities, the cell potentials are diminished for the entire range of temperatures investigated. The trend for increasing current densities

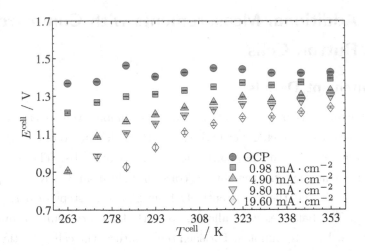

Figure B.1.: Averaged cell potentials of three commercial zinc air battery button cells from GP at various temperatures and set currents. The standard deviation for the three batteries is indicated as errorbar.

is in line with the polarization curve measurements with the in-house set-up shown in figure 5.1 (a) at 298 K, and reflects the typical polarization behavior of ZABs as presented in subchapter 1.1 and figure 1.2.

The effect of temperature on the polarization behavior is less distinct for elevated temperatures than for lower temperatures: Changing the current density from $0.98 \, \text{mA} \cdot \text{cm}^{-2}$ to $19.60 \, \text{mA} \cdot \text{cm}^{-2}$ at 283 K and 353 K, evokes an offset in cell potential of 0.54 V and 0.20 V, respectively. This can be explained by the enhanced diffusion of oxygen into the reaction zone at the CL at 353 K in comparison to 283 K. At elevated temperatures, a greater oxygen concentration is present at the CL and thus a lower overpotential at the air electrode emerges, which increases the cell potential obtained. In addition, all reaction kinetics at zinc electrode and air electrode are enhanced at elevated temperatures. Moreover, the ionic conductivity of the electrolyte is enhanced at elevated temperatures. The sum of these effects influences the cell potential in a positive way for higher temperatures and increased current densities. All aforementioned findings are in line with findings for ZAB button cells of a different type and manufacturer (see [29], p. 13.29, figure 13.28).

The results imply for ZABs that better polarization performance can be obtained at high temperatures. This holds especially at higher current densities. However, no conclusion can be drawn for the long-term stable discharge and charge of ZABs at this point. Negative effects, such as evaporation of water in the air electrode (see chapter 7), might be enhanced and can possibly cause an earlier battery failure at elevated temperatures.

The impact of the operating temperature on ZAB operation is not included in the model-based analysis of this thesis. However, the here presented results can be used to estimate the temperature impact on the cell potential for ZABs obtained by the simulations in this thesis.

Temperature Impact on Impedance Spectra

Figure B.2 depicts the Nyquist-plots of the impedance response of one commercially available button cell ZAB. It was operated consecutively at 298 K, 308 K, and 318 K, respectively, in a climate chamber with passive air supply to reveal the temperature impact on the high frequency and low frequency region of the impedance response. Both, the pronounced arc in the low frequency region and the less pronounced one in the high frequency region, appear as in the EIS measurement results shown before. Their deviation in the low frequency region is only approximately $1 \, \Omega \cdot cm^2$ of Z_{imag} in both frequency regions. However, it can be observed that the size of the arcs depicted is decreasing with decreasing temperature. The trend observed for the arc in the low frequency region can be explained by the enhanced reaction kinetics at both electrodes at elevated temperatures. In addition, the diffusive transport of oxygen into the CL of the air electrode might be enhanced at higher temperatures, which might possibly contribute to lower impedance values at the higher temperatures shown.

The enlargement in figure B.2 shows the impedance response in the high frequency region. The area specific ohmic resistance is $0.4862\ \Omega \cdot cm^2$ for the measurement at 298 K, $0.4351\ \Omega \cdot cm^2$, and $0.3995\ \Omega \cdot cm^2$ for 318 K. The trend observed for the impedance in the high frequency region might be predominantly due to an enhanced ionic conductivity of the electrolyte at elevated temperatures.

The results imply for the operation of ZABs that the reaction and transport processes at both electrodes are affected positively by elevated temperatures, so that consequently a better ZAB performance can be expected when operating at higher temperatures. Again, this result does not allow for predictions for the electrically rechargeable ZAB. A compromise between high temperature operation and long-term stable operation during discharge and charge cycles has to be elucidated by considering the air-composition impact on the half-open ZAB (see chapter 9).

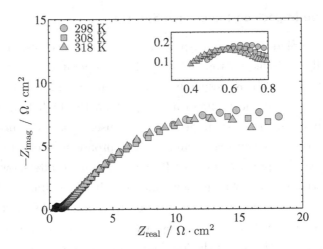

Figure B.2.: Nyquist-plot for EIS results for a GP button cell at different temperatures.

References

[1] E. Linden and T. B. Reddy. *Handbook of Batteries*. McGraw-Hill, New York, 3rd edition, 2001.

[2] U. Krewer. Portable Energiesysteme: Von elektrochemischer Wandlung bis Energy Harvesting. *Chemie Ingenieur Technik*, 83(11):1974–1983, 2011.

[3] A. A. Mohamad. Zn/gelled 6 M KOH/O_2 Zinc–Air Battery. *Journal of Power Sources*, 159(1):752–757, 2006.

[4] S. Müller, F. Holzer, and O. Haas. Optimized Zinc Electrode for the Rechargeable Zinc-Air Battery. *Journal of Applied Elechtrochemistry*, 28:895–898, 1998.

[5] A. G. Briggs, N. A. Hampson, and A. Marshall. Concentrated Potassium Zincate Solutions Studied Using Laser Raman Spectroscopy and Potentiometry. *Journal of the Chemical Society*, 70:1978–1990, 1974.

[6] B. Beverskog and I. Puigdomenech. Revised Pourbaix Diagrams for Zinc at 25-300 Degrees Celsius. *Corrosion Science*, 39(1):107–114, 1997.

[7] J. Lee, S. Tai Kim, R. Cao, N. Choi, M. Liu, K. T. Lee, and J. Cho. Metal-Air Batteries with High Energy Density: Li-Air versus Zn-Air. *Advanced Energy Materials*, 1(1):34–50, 2011.

[8] O. Haas, S. Müller, and K. Wiesener. Wiederaufladbare Zink/Luftsauerstoff-Batterien. *Chemie Ingenieur Technik*, 68(5):524–542, 1996.

[9] W. G. Sunu and D. N. Bennion. Transient and Failure Analyses of the Porous Zinc Electrode. *Journal of the Electrochemical Society*, 127(9):2017–2025, 1980.

[10] L. Jörissen. Bifunctional Oxygen/Air Electrodes. *Journal of Power Sources*, 155(1):23–32, 2006.

[11] K. Kinoshita. *Electrochemical Oxygen Technology*. John Wiley & Sons, Danvers, 1992.

[12] S. Müller, K. Striebel, and O. Haas. $La_{0.6}Ca_{0.4}CoO_3$: A Stable and Powerful Catalyst for Bifunctional Air Electrodes. *Electrochimica Acta*, 39(11/12):1661–1668, 1994.

[13] I. Krej, P. Van, and A. Trojnek. Transport of $Zn(OH)_4^{2-}$-Ions across a Polyolefin Microporous Membrane. *Journal of the Electrochemical Society*, 140(8):2–6, 1993.

[14] P. Arora and Z. Zhang. Battery Separators. *Chemical Reviews*, 104(10):4419–4462, 2004.

[15] E. L. Cussler. *Diffusion: Mass Transfer in Fluid Systems*. Cambridge University Press, New York, 3rd edition, 2007.

[16] Y. Li and H. Dai. Recent Advances in Zinc-Air Batteries. *Chemical Society Reviews*, 43:5257–5275, 2014.

[17] R. Gilliam, J. Graydon, D. Kirk, and S. Thorpe. A Review of Specific Conductivities of Potassium Hydroxide Solutions for Various Concentrations and Temperatures. *International Journal of Hydrogen Energy*, 32(3):359–364, 2007.

[18] M. B. Liu, B. R. Faulds, G. M. Cook, and N. P. Yao. Conductivity of KOH Electrolyte Supersaturated with Zincate. *Journal of the Electrochemical Society*, 128(10):2049–2052, 1981.

[19] D. M. See and R. E. White. Temperature and Concentration Dependence of the Specific Conductivity of Concentrated Solutions of Potassium Hydroxide. *Journal of Chemical and Engineering Data*, 42(6):1266–1268, 1997.

[20] K. Harting, U. Kunz, and T. Turek. Zinc-Air Batteries: Prospects and Challenges for Future Improvement. *Zeitschrift für Physikalische Chemie*, 226(2):151–166, 2012.

[21] D. U. Lee, J. Scott, H. W. Park, J. Choi, and Z. Chen. Morphologically Controlled Co_3O_4 Nanodisks as Practical Bi-Functional Catalyst for Rechargeable Zinc-Air Battery Applications. *Electrochemistry Communications*, 43:109–112, 2014.

[22] J. F. Parker, C. N. Chervin, and E.S. Nelson. Wiring Zinc in Three Dimensions Re-Writes Battery Performance-Dendrite-Free Cycling. *Energy and Environmental Science*, 7:1117–1124, 2014.

[23] W. Walker and F. R. Wilkins. Primary Battery. U.S. patent 524.229 and 524.291, 1894.

[24] G. W. Heise and E. A. Schumacher. Dry Cell of the Flat Type. U.S. patent 1890178 A, 1932.

[25] G. W. Heise and E. A. Schumacher. An Air-Depolarized Primary Cell with Caustic Alkali Electrolyte. *Journal of the Electrochemical Society*, 62(1):383–391, 1932.

[26] J. E. Schmidt. Air-Depolarized Primary Battery. U.S. patent 3788900 A, 1974.

[27] A. Marshall, N. A. Hampson, and J. S. Drury. The Discharge Behaviour of the Zinc/Air Slurry Cell. *Journal of Electroanalytical Chemistry and Interfacial Electrochemistry*, 59(1):33–40, 1975.

[28] J. Goldstein, I. Brown, and B. Koretz. New Developments in the Electric Fuel Ltd. Zinc Air System. *Journal of Power Sources*, 80(1–2):171–179, 1999.

[29] J. Passaniti, D. Carpenter, and R. Mckenzie. Button Cell Batteries: Silver Oxide–Zinc and Zinc-Air Systems. In E. Linden and T. B. Reddy, editors, *Handbook of Batteries*, chapter 13, pages 13.1–13.28. McGraw-Hill, New York, 2011.

[30] V. Caramia and B. Bozzini. Materials Science Aspects of Zinc Air Batteries: a Review. *Materials for Renewable and Sustainable Energy*, 3(2):1–12, 2014.

[31] P. Pei, K. Wang, and Z. Ma. Technologies for Extending Zinc–Air Battery's Cyclelife: a Review. *Applied Energy Materials*, 128:315–324, 2014.

[32] P. Pei, Z. Ma, K. Wang, X. Wang, M. Song, and H. Xu. High Performance Zinc Air Fuel Cell Stack. *Journal of Power Sources*, 249:13–20, 2014.

[33] H. Yang, Y. Cao, X. Ai, and L. Xiao. Improved Discharge Capacity and Suppressed Surface Passivation of Zinc Anode in Dilute Alkaline Solution Using Surfactant Additives. *Journal of Power Sources*, 128(1):97–101, 2004.

[34] M. N. Masri and A. A. Mohamad. Effect of Adding Potassium Hydroxide to an Agar Binder for Use as the Anode in Zn–Air Batteries. *Corrosion Science*, 51(12):3025–3029, 2009.

[35] J. F. Drillet, M. Adam, S. Barg, A. Herter, D. Koch, V. Schmidt, and M. Wilhelm. Development of a Novel Zinc/Air Fuel Cell with a Zn Foam Anode, a PVA/KOH Membrane and a MnO/SiOC-Based Air Cathode. *ECS Transactions*, 28(32):13–24, 2010.

[36] M. N. I. Masri, M. F. N. Nazeri, C. Y. Ng, and A. A. Mohamad. Tapioca Binder for Porous Zinc Anodes Electrode in Zinc–Air Batteries. *Journal of King Saud University - Engineering Sciences*, In Press, 2013.

[37] Y. Takeshita, S. Fujimoto, and M. Sudoh. Design of Rechargeable Air Diffusion Cathode for Metal-Air Battery Using Alkaline Solution. *ECS Transactions*, 50(19):3–12, 2013.

[38] H. Ma, B. Wang, Y. Fan, and W. Hong. Development and Characterization of an Electrically Rechargeable Zinc-Air Battery Stack. *Energies*, 7(10):6549–6557, 2014.

[39] D. Phihusut, J. D. Ocon, B. Jeong, J. W. Kim, J. K. Lee, and J. Lee. Gently Reduced Graphene Oxide Incorporated into Cobalt Oxalate Rods as Bifunctional Oxygen Electrocatalyst. *Electrochimica Acta*, 140:404–411, 2014.

[40] X. An, D. Shin, J. D. Ocon, J. K. Lee, Y. Son, and J. Lee. Electrocatalytic Oxygen Evolution Reaction at a FeNi Composite on a Carbon Nanofiber Matrix in Alkaline Media. *Chinese Journal of Catalysis*, 35(6):3–7, 2014.

[41] J. Ng, M. Tang, and T. F. Jaramillo. A Carbon-Free, Precious-Metal-Free, High-Performance O_2 Electrode for Regenerative Fuel Cells and Metal–Air Batteries. *Energy and Environmental Science*, 7:2017–2024, 2014.

[42] W. G. Hardin, J. T. Mefford, D. A. Slanac, B. B. Patel, X. Wang, S. Dai, X. Zhao, R. S. Ruoff, K. P. Johnston, and K. J. Stevenson. Tuning the Electrocatalytic Activity of Perovskites Through Active Site Variation and Support Interactions. *Chemistry of Materials*, 26(11):3368–3376, 2014.

[43] Y. Li, M. Gong, Y. Liang, J. Feng, J. Kim, H. Wang, G. Hong, B. Zhang, and H. Dai. Advanced Zinc-Air Batteries Based on

High-Performance Hybrid Electrocatalysts. *Nature Communications*, 4(1805), 2013.

[44] D. R. Rolison, J. F. Parker, and J. W. Long. Zinc Electrodes for Batteries. U.S. patent 20140147757 A1, 2014.

[45] P. Gehrke. Personal Communication. (Grillo-Werke Aktiengesellschaft, Goslar, Germany), December 2013.

[46] A. Bard and L. Faulkner. *Electrochemical Methods: Fundamentals and Applications.* John Wiley & Sons, New York, 2nd edition, 2001.

[47] J. Larminie and A. Dicks. *Fuel Cell Systems Explained.* John Wiley & Sons, Chichester, 2003.

[48] E. Barsoukov and J. R. Macdonald. *Impedance Spectroscopy: Theory, Experiment, and Applications.* John Wiley & Sons, New York, 2005.

[49] F. Huet. A Review of Impedance Measurements for Determination of the State-of-Charge or State-of-Health of Secondary Batteries. *Journal of Power Sources*, 70(1):59–69, 1998.

[50] J. Banhart. *Advanced Tomographic Methods in Materials Research and Engineering.* Oxford University Press, New York, 2008.

[51] P. Sprawls. *Physical Principles of Medical Imaging.* Aspen Publishers, Gaithersburg, 2nd edition, 1993.

[52] T. Arlt, I. Manke, K. Wippermann, H. Riesemeier, J. Mergel, and J. Banhart. Investigation of the Local Catalyst Distribution in an Aged Direct Methanol Fuel Cell MEA by Means of Differential Synchrotron X-ray Absorption Edge Imaging with High Energy Resolution. *Journal of Power Sources*, 221:210–216, 2013.

[53] H. Markötter, I. Manke, J. Haussmann, and T. Arlt. Combined Synchrotron X-ray Radiography and Tomography Study of Water Transport in Gas Diffusion Layers. *Micro and Nano Letters*, 7(7):689–692, 2012.

[54] Y. Waseda, E. Matsubara, and K. Shinoda. *Fundamental Properties of X-rays*. Springer-Verlag, Berlin/Heidelberg, 2011.

[55] J. H. Hubbell and S. M. Seltzer. Tables of X-Ray Mass Attenuation Coefficients and Mass Energy-Absorption Coefficients from 1 keV to 20 MeV for Elements Z equal 1 to 92 and 48 Additional Substances of Dosimetric Interest, 1996. NISTIR 5632, Physical Reference Data.

[56] M. König. Einfluss der Umgebungsluftfeuchte auf das Betriebsverhalten von Zink-Luftsauerstoff-Batterien. Studienarbeit, Betreuer: D. Schröder, Gutachter: U. Krewer, TU Braunschweig, Fakultät für Maschinenbau, Institut für Energie- und Systemverfahrenstechnik, 2013.

[57] D. Schröder, N. N. Sinai Borker, M. König, and U. Krewer. Performance of Zinc Air Batteries with Added K_2CO_3 in the Alkaline Electrolyte. *Journal of Applied Electrochemistry*, 45(5):427–437, 2015.

[58] T. Arlt. Personal Communication. (Helmholtz-Zentrum Berlin, Institut Angewandte Materialforschung, Berlin, Germany). June 2014.

[59] D. Schröder, T. Arlt, U. Krewer, and I. Manke. Analyzing Transport Paths in the Air Electrode of a Zinc Air Battery using X-ray Tomography. *Electrochemistry Communications*, 40:88–91, 2014.

[60] T. Arlt, D. Schröder, U. Krewer, and I. Manke. In Operando Monitoring of State of Charge and Species Distribution in Zinc Air Batteries with X-ray Tomography and Model-Based Analysis. *Physical Chemistry Chemical Physics*, 16:22273–22280, 2014.

[61] D. Schröder, M. König, and U. Krewer. Experimental EIS Analysis of Environmental Impact on Zinc Air Battery Operation. *Meeting Abstract, 223rd ECS Meeting, Toronto, Canada*, MA2013-01 364(6):364, 2013.

[62] W. S. Rasband. ImageJ. U.S. National Institutes of Health, Bethesda, Maryland, USA, imagej.nih.gov/ij/, ©1997-2012.

[63] D. Tromans. Oxygen Solubility Modeling in Inorganic Solutions: Concentration, Temperature and Pressure Effects. *Hydrometallurgy*, 50(3):279–296, 1998.

[64] R. De Levie. On Porous Electrodes in Electrolyte Solutions-IV. *Electrochimica Acta*, 9(9):1231–1245, 1964.

[65] C. Cachet and R. Wiart. The Pore Texture of Zinc Electrodes characterized by Impedance Measurements. *Electrochimica Acta*, 29(2):145–149, 1984.

[66] J. Hendrikx, W. Visscher, and E. Barendrecht. Impedance Measurements of Zinc and Amalgamated Zinc Electrodes in Alkaline Electrolyte. *Electrochimica Acta*, 30(8):999–1006, 1985.

[67] C. Cachet, B. Saidani, and R. Wiart. The Behavior of Zinc Electrode in Alkaline Electrolytes. *Journal of the Electrochemical Society*, 138(3):644–654, 1991.

[68] C. Cachet, U. Ströder, and R. Wiart. The Kinetics of Zinc Electrode in Alkaline Zincate Electrolytes. *Electrochimica Acta*, 27(7):903–908, 1982.

[69] J. Suntivich, H. A. Gasteiger, N. Yabuuchi, and Y. Shao-Horn. Electrocatalytic Measurement Methodology of Oxide Catalysts Using a Thin-Film Rotating Disk Electrode. *Journal of the Electrochemical Society*, 157(8):B1263–B1268, 2010.

[70] J. Suntivich, H. A. Gasteiger, N. Yabuuchi, H. Nakanishi, J. B. Goodenough, and Y. Shao-Horn. Design Principles for Oxygen-Reduction Activity on Perovskite Oxide Catalysts for Fuel Cells and Metal-Air Batteries. *Nature Chemistry*, 3(7):546–550, 2011.

[71] E. Deiss, F. Holzer, and O. Haas. Modeling of an Electrically Rechargeable Alkaline Zn-Air Battery. *Electrochimica Acta*, 47(25):3995–4010, 2002.

[72] V. Laue. Modellierung und Simulation des Stofftransports durch die Luftelektrode einer Zink-Luft-Batterie. Bachelorarbeit, Betreuer: D. Schröder, Gutachter: U. Krewer, TU Braunschweig, Fakultät für Maschinenbau, Institut für Energie- und Systemverfahrenstechnik, 2013.

[73] D. Schröder, V. Laue, and U. Krewer. Pulse-Current-Operation of Alkaline Gas-Diffusion-Electrodes. *Meeting Abstract, ACOMEN- 6th International Conference on Advanced Computational Methods in Engineering, Ghent, Belgium*, 2014.

[74] D. Schröder, V. Laue, and U. Krewer. Numerical Simulation of Gas-Diffusion-Electrodes with Moving Gas-Liquid Interface: A Study on Pulse-Current Operation and Electrode Flooding. *Computers & Chemical Engineering*, 84:217–225, 2016.

[75] C. Y. Wang, W. B. Gu, and B. Y. Liaw. Micro-Macroscopic Coupled Modeling of Batteries and Fuel Cells I. Model Development. *Journal of the Electrochemical Society*, 145(10):3407–3417, 1998.

[76] R. E. Davis, G. L. Horvath, and C. W. Tobias. The Solubility and Diffusion Coefficient of Oxygen in Potassium Hydroxide Solutions. *Electrochimica Acta*, 12(3):287–297, 1967.

[77] R. C. Weast, M. J. Astle, and W. H. Beyer. *CRC Handbook of Chemistry and Physics*. CRC Press, Boca Raton, 62th edition, 1982.

[78] L. F. Shampine and M. W. Reichelt. The Matlab ODE Suite. *SIAM Journal on Scientific Computing*, 18(1):1–22, 1997.

[79] N. Shaigan, W. Qu, and T. Takeda. Morphology Control of Electrodeposited Zinc from Alkaline Zincate Solutions for Rechargeable Zinc Air Batteries. *ECS Transactions*, 28(32):35–44, 2010.

[80] J. Balej. Water Vapour Partial Pressures and Water Activities in Potassium and Sodium Hydroxide Solutions Over Wide Concentration and Temperature Ranges. *International Journal of Hydrogen Energy*, 10(4):233–243, 1985.

[81] D. R. Stull. Vapor Pressure of Pure Substances Organic Compounds. *Industrial and Engineering Chemistry*, 39:517–540, 1947.

[82] L. Kucka, E. Y. Kenig, and A. Górak. Kinetics of the Gas-Liquid Reaction Between Carbon Dioxide and Hydroxide Ions. *Industrial and Engineering Chemistry Research*, 41(24):5952–5957, 2002.

[83] J. F. Drillet, F. Holzer, T. Kallis, S. Müller, and V. M. Schmidt. Influence of CO_2 on the Stability of Bifunctional Oxygen Electrodes for Rechargeable Zinc/Air Batteries and Study of Different CO_2 Filter Materials. *Physical Chemistry Chemical Physics*, 3:368–371, 2001.

[84] A. Renuka, A. Veluchamy, and N. Venkatakrishnan. Effect of Carbonate Ions on the Behaviour of Zinc in 30% KOH. *Journal of Power Sources*, 34:381–385, 1991.

[85] D. Schröder and U. Krewer. Model Based Quantification of Air-Composition Impact on Secondary Zinc Air Batteries. *Electrochimica Acta*, 117:541–553, 2014.

[86] P. C. Tseng and W. S. Ho. Carbon Dioxide Absorption into Promoted Carbonate Solutions. *AIChE Journal*, 34(6):922–931, 1988.

[87] R. Pohorecki and W. Moniuk. Kinetics of Reaction Between Carbon Dioxide and Hydroxyl Ions in Aqueous Electrolyte Solutions. *Chemical Engineering Science*, 43(7):1677–1684, 1988.

[88] E. Gileadi. *Electrode Kinetics for Chemists, Chemical Engineers and Materials Scientists*. VCH Publishers, New York, 1993.

[89] P. Bro and H.Y. Kang. The Low-Temperature Activity of Water in Concentrated KOH Solutions. *Journal of the Electrochemical Society*, 118(9):1430–1434, 1971.

[90] U. Krewer, D. Schröder, and A. Ramirez-Sanchez. Alternative Battery Technologies, their Potential and Challenges. *Meeting Abstract, 10th Symposium on Hybrid and Electric Vehicles (HEV), Braunschweig, Germany*, 2013.

[91] W. M. Haynes. *CRC Handbook of Chemistry and Physics - Aqueous Solubility of Inorganic Compounds at Various Temperatures*. CRC Press, Boca Raton, 94th edition, 2013.

[92] Z. Peng, S. Freunberger, Y. Chen, and P. G. Bruce. A Reversible and Higher-Rate Li-O_2 Battery. *Science*, 337(6094):563–566, 2012.

[93] J. Newman and C. W. Tobias. Theoretical Analysis of Current Distribution in Porous Electrodes. *Journal of the Electrochemical Society*, 109(12):1183–1191, 1962.

[94] E. J. Podlaha, H. G. Delighanni, and G. Stafford. Electrodeposition Fueled by Newman and Tobias. *The Electrochemical Society - Interface*, Spring Volume:39–42, 2010.

[95] M. Doyle, T. F. Fuller, and J. Newman. Modeling of Galvanostatic Charge and Discharge of the Lithium/Polymer/Insertion Cell. *Journal of the Electrochemical Society*, 140(6):1526–1533, 1993.

[96] E. Deiss, D. Häringer, P. Novak, and O. Haas. Modeling of the Charge–Discharge Dynamics of Lithium Manganese Oxide Electrodes for Lithium-Ion Batteries. *Electrochimica Acta*, 46:4185–4196, 2001.

[97] U. Sahapatsombut, H. Cheng, and K. Scott. Modelling of Operation of a Lithium-Air Battery with Ambient Air and Oxygen-Selective Membrane. *Journal of Power Sources*, 249:418–430, 2014.

[98] J. Newman and K. E. Thomas-Alyea. *Electrochemical Systems*. Wiley-Interscience, Hoboken, 3rd edition, 2004.

[99] Z. Mao and R. E. White. Mathematical Modeling of a Primary Zinc/Air Battery. *Journal of the Electrochemical Society*, 139(4):1105–1114, 1992.

[100] J. Johansen, T. Farrell, and C. Please. Modelling of Primary Alkaline Battery Cathodes: A Simplified Model. *Journal of Power Sources*, 156(2):645–654, 2006.

[101] J. Chen and H. Y. Che. Modeling of Cylindrical Alkaline Cells, IV. Dissolution-Precipitation Model for the Anode. *Journal of the Electrochemical Society*, 140(5):1213–1218, 1993.

[102] K. W. Choi, D. N. Bennion, and J. Newman. Engineering Analysis of Shape Change in Zinc Secondary Electrodes. *Journal of the Electrochemical Society*, 123(11):1628–1637, 1976.

[103] G. G. Kumar and S. Sampath. Electrochemical Characterization of a Zinc-Based Gel-Polymer Electrolyte and its Application in Rechargeable Batteries. *Journal of the Electrochemical Society*, 150(5):A608–A615, 2003.

[104] F. Buker, C. Müller, T. Hanemann, D. Hertkorn, and H. Reinecke. Modeling of the Electrical Properties of Bidirectional Alkaline Air Electrodes. *Journal of the Electrochemical Society*, 161(6):A1019–A1022, 2014.

[105] Y. Ko and S. Park. Zinc Oxidation in Dilute Alkaline Solutions Studied by Real-Time Electrochemical Impedance Spectroscopy. *The Journal of Physical Chemistry C*, 116(13):7260–7268, 2012.

[106] P. W. C. Northrop, B. Suthar, V. Ramadesigan, S. Santhanagopalan, R. D. Braatz, and V. R. Subramanian. Efficient Simulation and Reformulation of Lithium-Ion Battery Models for Enabling Electric

Transportation. *Journal of the Electrochemical Society*, 161(8):E3149–E3157, 2014.

[107] S. Siahrostami, V. Tripkovic, K. T. Lundgaard, K. E. Jensen, H. Hansen, J.S. Hummelshoj, J. S. G. Myrdal, T. Vegge, J. K. Norskov, and J. Rossmeisl. First Principles Investigation of Zinc-Anode Dissolution in Zinc-Air Batteries. *Physical Chemistry Chemical Physics*, 15(17):6416–21, 2013.

[108] R. D. Braatz. Multiscale Modeling and Design of Electrochemical Systems. In R. C. Alkire, editor, *Advances in Electrochemical Science and Engineering*, volume 10, chapter 4, pages 298–334. Wiley-VCH, 2008.

[109] M. Jain and J. W. Weidner. Material Balance Modification in One-Dimensional Modeling of Porous Electrodes. *Journal of the Electrochemical Society*, 146(4):1370–1374, 1999.

[110] M. C. Kimble and R. E. White. A Mathematical Model of a Hydrogen/Oxygen Alkaline Fuel Cell. *Journal of the Electrochemical Society*, 138(11):3370–3382, 1991.

[111] U. Krewer, D. Schröder, and C. Weinzierl. Scenario-Based Analysis of Potential and Constraints of Alkaline Electrochemical Cells. *Computer Aided Chemical Engineering*, 33(24th European Symposium on Computer Aided Process Engineering):1237–1242, 2014.

[112] O. Levenspiel. *Chemical Reaction Engineering*. John Wiley & Sons, Danvers, 3rd edition, 1999.

[113] R. Othman, W. J. Basirun, and A. H. Yahaya. Hydroponics Gel as a New Electrolyte Gelling Agent for Alkaline Zinc-Air Cells. *Journal of Power Sources*, 103:34–41, 2001.

[114] R. Othman and A. H. Yahaya. A Zinc-Air Cell Employing a Porous Zinc Electrode Fabricated from Zinc-Graphite-Natural

Biodegradable Polymer Paste. *Journal of Applied Electrochemistry*, 32(12):1347–1353, 2002.

[115] M. Hilder, B. Winther-Jensen, and N. sB. Clark. The Effect of Binder and Electrolyte on the Performance of Thin Zinc-Air Battery. *Electrochimica Acta*, 69:308–314, 2012.

[116] T. Yamanaka, T. Takeguchi, H. Takahashi, and W. Ueda. Water Transport during Ion Conduction in Anion-Exchange and Cation-Exchange Membranes. *Journal of the Electrochemical Society*, 156(7):B831–B835, 2009.

[117] C. Weinzierl and U. Krewer. Model-based Analysis of Water Management in Alkaline Direct Methanol Fuel Cells. *Journal of Power Sources*, 268:911–921, 2014.

[118] J. F. Zemaitis, D. M. Clark, M. Rafal, and N. D. Scrivner. *Handbook of Aqueous Electrolyte Thermodynamics - Theory and Application*. Wiley-Interscience and the American Institute of Chemical Engineers, New York, 1986.

[119] W. J. Hamer and Y. Wu. Osmotic Coefficients and Mean Activity Coefficients of Uni-univalent Electrolytes in Water at 25 degrees Celsius. *Journal of Physical and Chemical Reference Data*, 1(4):1047–1100, 1972.

[120] S. T. P. Psaltis and T. W. Farrell. Comparing Charge Transport Predictions for a Ternary Electrolyte Using the Maxwell–Stefan and Nernst–Planck Equations. *Journal of the Electrochemical Society*, 158(1):A33–A42, 2011.

[121] W. Dreyer, C. Guhlke, and R. Müller. Overcoming the Shortcomings of the Nernst-Planck Model. *Physical Chemistry Chemical Physics*, 15(19):7075–7086, 2013.

[122] J. Fuhrmann. Comparison and Numerical Treatment of Generalised Nernst–Planck Models. WIAS Preprint No. 1940, http://www.wias-berlin.de/preprint/1940/wias_preprints_1940.pdf, last accessed 14th of February 2015, 2014.

[123] J. Fuhrmann. Personal Communication. (Weierstrass Institut, Forschungsgruppe Numerische Mathematik und Wissenschaftliches Rechnen, Berlin, Germany), June 2014.

[124] J. J. Kriegsmann and H. Y. Cheh. A Binary Electrolyte Model of a Cylindrical Alkaline Cell. *Journal of Power Sources*, 85(2):190–202, 2000.

[125] Z. Guo, X. Dong, S. Yuan, Y. Wang, and Y. Xia. Humidity Effect on Electrochemical Performance of $Li-O_2$ Batteries. *Journal of Power Sources*, 264:1–7, 2014.

[126] I. Demirdžić and S. Muzaferija. Numerical method for coupled fluid flow, heat transfer and stress analysis using unstructured moving meshes with cells of arbitrary topology. *Computer Methods in Applied Mechanics and Engineering*, 125(1):235–255, 1995.

[127] R. Eymard, T. Gallouët, and R. Herbin. *Handbook of Numerical Analysis*, volume 7, chapter 'Finite Volume Methods', pages 713–1018. North-Holland Publishing Company, Amsterdam, 2000.

[128] J. H. Ferziger and M. Perić. *Computational Methods for Fluid Dynamics*. Springer-Verlag, Berlin, 3rd edition, 2002.

[129] G. Zhou, L. Chen, and J. P. Seaba. Effects of Property Variation and Ideal Solution Assumption on the Calculation of the Limiting Current Density Condition of Alkaline Fuel Cells. *Journal of Power Sources*, 196(11):4923–4933, 2011.

[130] J. Haubrock, J. Hogendoorn, and G. Versteeg. The Applicability of Activities in Kinetic Expressions: A More Fundamental Approach

to Represent the Kinetics of the System CO_2-OH-Salt in Terms of Activities. *Chemical Engineering Science*, 62(21):5753–5769, 2007.

[131] Q. Liu and R. A. Lange. New Density Measurements on Carbonate Liquids and the Partial Molar Volume of the $CaCO_3$ Component. *Contributions to Mineralogy and Petrology*, 146(3):370–381, 2003.

[132] P. M. Mathias. Correlation for the Density of Multicomponent Aqueous Electrolytes. *Industrial and Engineering Chemistry Research*, 43(19):6247–6252, 2004.

[133] T. P. Dirkse, D. De Wit, and R. Shoemaker. The Anodic Behavior of Zinc in KOH Solutions. *Journal of the Electrochemical Society*, 115(5):442–444, 1968.

[134] J. O. M. Bockris and T. Otagawa. The Electrocatalysis of Oxygen Evolution on Perovskites. *Journal of the Electrochemical Society*, 131(2):290–302, 1984.

[135] K.N. Grew, X. Ren, and D. Chu. Effects of Temperature and Carbon Dioxide on Anion Exchange Membrane Conductivity. *Electrochemical and Solid-State Letters*, 14(12):B127–B131, 2011.

[136] I. Arise, S. Kawai, Y. Fukunaka, and F.R. McLarnon. Numerical Calculation of Ionic Mass-Transfer Rates Accompanying Anodic Zinc Dissolution in Alkaline Solution. *Journal of the Electrochemical Society*, 157(2):A171, 2010.

[137] R. B. Bird, Stewart W. E., and E. N. Lightfoot. *Transport Phenomena*. John Wiley & Sons, Danvers, 2nd edition, 2002.

[138] J. Yue, G. Chen, Q. Yuan, L. Luo, and Y. Gonthier. Hydrodynamics and Mass Transfer Characteristics in Gas–Liquid Flow Through a Rectangular Microchannel. *Chemical Engineering Science*, 62(7):2096–2108, 2007.

[139] M. Laliberte and W. E. Cooper. Model for Calculating the Density of Aqueous Electrolyte Solutions. *Journal of Chemical & Engineering Data*, 49(5):1141–1151, 2004.

[140] G. S. Kell. Density, Thermal Expansivity, and Compressibility of Liquid Water from 0 to 150 Degrees. Correlations and Tables for Atmospheric Pressure and Saturation Reviewed and Expressed on 1968 Temperature Scale. *Journal of Chemical and Engineering Data*, 20(1):97–105, 1975.

[141] G. H. Jeffery, J. Bassett, J. Mendham, and R. C. Denney. *Vogel's Textbook of Quantitative Chemical Analysis*. Longman Scientific & Technical, Harlow (Essex), 5th edition, 1989.

Printed in the United States
by Baker & Taylor Publisher Services